Modernity and Technology

Postphenomenology and the Philosophy of Technology

Editor-in-Chief: Robert Rosenberger

Executive Editors: Peter-Paul Verbeek and Don Ihde†

As technologies continue to advance, they correspondingly continue to make fundamental changes to our lives. Technological changes have effects on everything from our understandings of ethics, politics, and communication, to gender, science, and selfhood. Philosophical reflection on technology can help draw out and analyze the nature of these changes, and help us to understand both the broad patterns of technological effects and the concrete details. The purpose of this series is to provide a publication outlet for the field of philosophy of technology in general, and the school of thought called "postphenomenology" in particular. The field of philosophy of technology applies insights from the history of philosophy to current issues in technology and reflects on how technological developments change our understanding of philosophical issues. Postphenomenology is the name of an emerging research perspective used by a growing international and interdisciplinary group of scholars. This perspective utilizes insights from the philosophical tradition of phenomenology to analyze human relationships with technologies, and it also integrates philosophical commitments of the American pragmatist tradition of thought.

Recent Titles in This Series

Modernity and Technology: A Philosophical Investigation of Martin Heidegger and Bruno Latour by Søren Riis

Moral Hermeneutics and Technology: Making Moral Sense through Human-Technology-World Relations by Olya Kudina

Postphenomenology and Imaging: How to Read Technology, edited by Samantha J. Fried and Robert Rosenberger

Postphenomenology and Architecture: Human Technology Relations in the Built Environment, edited by Lars Botin and Inger Berling Hyams

Technology and Anarchy: A Reading of Our Era, by Simona Chiodo

How Scientific Instruments Speak: Postphenomenology and Technological Mediations in Neuroscientific Practice by Bas de Boer

Feedback Loops: Pragmatism about Science & Technology, edited by Andrew Wells Garnar and Ashley Shew

Sustainability in the Anthropocene: Philosophical Essays on Renewable Technologies, edited by Róisín Lally

Modernity and Technology

A Philosophical Investigation of Martin Heidegger and Bruno Latour

Søren Riis

LEXINGTON BOOKS
Lanham • Boulder • New York • London

Published by Lexington Books
An imprint of The Rowman & Littlefield Publishing Group, Inc.
4501 Forbes Boulevard, Suite 200, Lanham, Maryland 20706
www.rowman.com

86-90 Paul Street, London EC2A 4NE

Copyright © 2025 by The Rowman & Littlefield Publishing Group, Inc.

All rights reserved. No part of this book may be reproduced in any form or by any electronic or mechanical means, including information storage and retrieval systems, without written permission from the publisher, except by a reviewer who may quote passages in a review.

British Library Cataloguing in Publication Information Available

Library of Congress Cataloging-in-Publication Data Available

Includes bibliographic references and index.
ISBN: 978-1-66692-957-7 (cloth)
ISBN: 978-1-66692-958-4 (electronic)

∞™ The paper used in this publication meets the minimum requirements of American National Standard for Information Sciences—Permanence of Paper for Printed Library Materials, ANSI/NISO Z39.48-1992.

Contents

Foreword *by Peter-Paul Verbeek*	vii
Introduction	1
1 Monsters of Modernity	13
2 Modernity in Action	29
3 Death and the End of Networks	43
4 Another Beginning	65
5 Meeting Half Way	91
6 The End of the World as We Know It	115
Bibliography	129
Index	137
About the Author	143

Foreword

by Peter-Paul Verbeek

Against the background of the many crises the world is currently going through as a result of technological developments, the idea that "Modernity" equals "improvement" seems further away than ever. Global warming, geopolitical tensions, and the disruptive cultural, societal, and political impacts of digital technologies have shown that technological developments can never be seen as a simple shortcut to paradise.

The debate around modernity has long been dominated by a tension between "conservatism" and "progressivism." While conservative thinkers reject modernity for the disenchantment, alienation, and inequality it would bring, progressive thinkers have emphasized how science and technology have great potential to address societal challenges and to improve the quality of life on Earth. This book brings together the ideas of two central and critical contributors to this debate: Martin Heidegger and Bruno Latour. And by doing so, it takes us beyond the binary opposition between merely being "for" or "against" modernity.

Both thinkers have raised controversy, albeit in unique and completely different ways. Martin Heidegger was not only a very influential philosopher but also a member of the Nazi party. He never distanced himself from the Nazis after the Second World War, and his personal notebooks, which were published decades after the war, contain explicitly anti-Semitic statements. All of this is a reason for some not to engage with his work anymore. Bruno Latour, on the other hand, was at the center of the so-called "Science Wars": a fierce debate about the alleged "relativism" of some thinkers in Science and Technology Studies (STS), who were accused of being "postmodernists" denying the existence of scientific facts and truth. In fact, though, Latour has always stayed away from postmodernism, labeling his own approach as "amodern" and claiming that we have actually "never been modern."

Also, between Heidegger and Latour was antagonism: Latour has explicitly distanced himself from Heidegger several times, criticizing Heidegger's nostalgic and anti-technological romanticism. But while Heidegger seemed to long for premodern times and Latour claimed that the idea of modernity is a big mistake, both thinkers have developed profound critiques of modernity—critiques that are more congenial than they seem to be at first sight. The "theoretical fieldwork" that Søren Riis undertook to study the relations between them lays bare many new insights that can provide guidance in addressing the crises we are currently facing.

Both Heidegger and Latour have developed new ways to understand the relations between humans and nature, subject and object, social sciences and natural sciences, human beings and technological artifacts, the living and the dead. Both have revealed how science and technology are connected to our understanding of the world and ourselves. The dialogue between their approaches, as brought about by this book, results in a new way to understand how modern science and technology require a new metaphysics that overcomes the split between human subjects and nonhuman objects, and a new philosophical anthropology that can do justice to the intricate relations between humans and technologies that have emerged with scientific and technological developments.

Modernity and Technology organizes a much-needed encounter between two thinkers who have never met but who have, each in their own way, profoundly shaped late twentieth and early twenty-first-century philosophy. Experiencing this encounter through the eyes of Søren Riis and against the background of the crises that the world is currently facing brings a new depth to our understanding of their work and to the ways in which we could address the challenges of the twenty-first century.

Introduction

Climate change, mass extinction, and surveillance capitalism are just a few of the major dangers facing late modernity. All these threats appear to be intensifying, and none of them would be possible without modern technologies. Humanity has unleashed astounding powers that have proven exceedingly difficult to control. Moreover, attempts at control have frequently resulted in the creation of further experimental technologies and exacerbated hazards. Without succumbing to fatalistic dystopian or technological determinist perspectives, there is an urgent need today to critically reexamine how we conceptualize and develop new technologies.

The French thinker Bruno Latour (1947–2022) comprehended both the systemic dangers posed by modern technologies and the vulnerability of life on Earth, but only in the last years leading up to his death. It wasn't until the 2010s and Latour was in his sixties that he really came to acknowledge the major systemic consequences of modern technology for life on Earth. In his last writings, he was a vocal critic of late Western civilization and its disastrous trajectory, without, however, explicitly admitting to problems pertaining to his prior long-standing optimism regarding technology (Latour 2017, 2018). Though his older German counterpart and critic of modernity, Martin Heidegger (1889–1976), did not live to witness these evolving global dangers, he nonetheless exhibited prophetic foresight in his warnings against modernity.

This book investigates the concept of modernity and its relationship with technology. This investigation is carried out by addressing the tension and mutual ground between two of the most prominent thinkers working on this highly debated relationship, Bruno Latour and Martin Heidegger. The phenomena and questions examined here range from the mundane to the complex, from the perilous to the aesthetic. It is posited that, despite apparent differences, these two authors exhibit remarkable and profound similarities when their respective understandings of modernity are scrutinized. Furthermore, I argue that these commonalities warrant consideration, as they may illuminate avenues of breakthrough and salvation from some of the major threats of the twenty-first century.

Two passages from Latour are taken as the point of departure for this potentially provocative book, since they may prove helpful in navigating the interpretive journey ahead. "Since nothing is, in and of itself, either equivalent, or not equivalent, two forces cannot associate without misunderstanding" (Latour 1993b, 168). The second illustrative yet profound passage is also taken from *The Pasteurization of France*. In this passage, Latour employs a phenomenological thought experiment to present a first-person perspective on the interpretive practice of science. This text is quoted here because it can also be related to the task of unfolding two remarkable thinkers such as Latour and Heidegger; it sets the stage for my interpretive endeavor:

> I don't know how things stand. I know neither who I am nor what I want, but others say they know on my behalf, others, who define me, link me up, make me speak, interpret what I say, and enroll me. Whether I am a storm, a rat, a rock, a lake, a lion, a child, a worker, a gene, a slave, the unconscious, or a virus, they whisper to me, they suggest, they impose an interpretation of what I am and what I could be. (Latour 1993b, 192)[1]

Heidegger and Latour have been seminal thinkers in different disciplinary domains for decades. Whereas Heidegger fundamentally influenced the course of philosophy in the twentieth century, Latour has been breaking new ground in sociology and science and technology studies for the last forty years, until he passed in 2022. Heidegger was primarily concerned with the basic principles of thinking and Being, the problems of modernity, and warned against the imperial disposition of modern technology, whereas Latour was renowned for his empirical investigations, fascination with technology, and claim that modernity is a mere illusion. The differences between the two thinkers are intensified by the fact that Latour, on several occasions, explicitly criticized Heidegger and his legacy, turning the apparent discrepancy into an open controversy (Latour 1993, 67, 1999, 176, 2005a, 23).[2]

Because Heidegger and Latour are mostly read by students and scholars from different disciplinary backgrounds, who often have progressive or conservative inclinations, these two remarkable thinkers are seldom studied in tandem. And when they are, it is most often in support of Latour's claims, which has further widened the gap between them. This book works in the opposite direction and claims that Heidegger and Latour are in fact much closer to each other than Latour is willing to admit, and, more importantly, it argues that associating Heidegger with Latour allows us to see fundamental similarities and interdisciplinary concerns and controversies between contemporary philosophy, sociology, and various appropriations of the past. This association of Latour and Heidegger not only advances our understanding of the intrinsic ways of being of humans and nonhumans, but also helps us see how common concerns can manifest themselves differently. Expressed less polemically—independently of any inspiration Latour might have drawn from Heidegger—reading these two major thinkers in conjunction gives significant and challenging insights into a range of contemporary issues being addressed by philosophy, sociology, and history alike.

From a biographical perspective, there are also a number of links relating Heidegger and Latour that are worthy of discussion in order to set the stage for the claims and arguments to follow. Heidegger and Latour grew up only a few hundred kilometers from each other in Catholic villages. Heidegger was born in Meßkirch, a small village on the border with France, and Latour in Beaune, Burgundy, which is relatively close to the *unbeaten tracks* of Heidegger in Germany. Both began their intellectual journeys studying theology and subsequently changed to philosophy (Ott 1992, 45f; Blok and Jensen 2012, 4f). Furthermore, Heidegger's first position as a professor of philosophy was in Marburg (1923–1927), Germany, where he worked closely with the theologian Rudolf Bultmann, who shared his hermeneutic approach to theology. As the prominent German theologian Eberhard Jüngel writes in his introduction to the publication of Bultmann and Heidegger's letter correspondence:

> From the beginning till the very end the friendship was filled with a heartfelt, exceptional mutual respect. For Heidegger, most important was the incorruptible judgment of his Marburgian friend. And Bultmann made, until old age, persistent effort to understand the profound thought of his Alemannic friend and to apply it to his own theological work. Karl Barth [famous German theologian] noted, not without irony, that during a meeting near Göttingen, Bultmann read aloud to him "for hours from his notes on Heidegger's lectures." (Bultmann 2009, VI; trans. Rebecca Walsh)

In the years following his graduation in philosophy, Latour emphasized time and again the decisive nature of Bultmann's early influence on his own development as a researcher (Latour 2010c, 600; Blok and Jensen 2012, vii). Both Latour and Heidegger developed profound critiques of modernity and shared a common interest in religion. Furthermore, during a lecture at Harvard, Latour talks about the contemporary German philosopher and Heidegger scholar Peter Sloterdijk, "I was born a Sloterdijkian."[3] Even readers not well-versed in mediation sociology may be intrigued by these common biographical traits. Hasty conclusions on their apparent differences may cover deeper commonalities between these two fascinating thinkers that are worthy of exploring.

In connection to the notion of modernity, this book emphasizes a series of shared matters of concern between Heidegger and Latour. It does not pretend to be comprehensive in the sense that it does not endeavor to address all of their common interests; rather, it strives to be selective and sensitive, as it focuses on only a few shared but significant topics and proceeds from there to establish a range of systematic associations and far-reaching links. In this sense, the book presents several connected examples of relational philosophy, which can be elaborated and serve as inspiration for further comparative philosophical studies.

First and foremost, the book addresses how modernity is understood by Latour and Heidegger. Their differing views of modernity allow us to see why they have traditionally reached very different readers. On the other hand, their differences manifest remarkable common interests, as many of Heidegger's and Latour's arguments and conclusions stand in a dubious inverse relation to each other, which calls for scrutiny. For example, Heidegger viewed modernity as a real and dangerous decline in history and as an epoch of Being that pushes human life to its very limits, whereas Latour argued for many years that modernity was an illusion and that especially the contemporary development of complex technologies was proof that there has never really been any accurate or trustworthy way to definitively delimit humans from nonhumans. Their respective understandings led Heidegger to develop his prominent notion of Dasein and Latour to unfold what he calls a symmetrical anthropology of humans and nonhumans.

Following these lines of thinking, an analysis of what it means to be human according to Latour and Heidegger shapes the second crucial concern of this book: philosophical anthropology. Latour and Heidegger agree on a number of fundamental characteristics of human beings; they both emphasize the intimate relation between humans and their surroundings, thus granting *things* an outstanding existential significance. In their respective ontologies, the hybridity of being human is critical, yet Heidegger does not see this as leading to a fundamental symmetry between humans and nonhumans as Latour does. Latour, who for years was preoccupied with science and technology studies, scientific truth production, and the role of humans and instruments in the laboratory, especially with respect to the shaping of the contemporary world picture, sees a fundamental symmetrical anthropology confirmed everywhere in the world of science and technologies. Heidegger, on the other hand, ultimately establishes a fundamental difference between humans and all other beings based on Dasein's own understanding of finitude. For these reasons in particular, neither thinker explicitly unfolds an ethical compass for humans. Latour is convinced that there is no specific moral compass only for human beings, as this would suggest an asymmetry between humans and nonhumans: there exists a common collective structure of being, which renders all beings responsive to each other, and this leads to an abstract and concrete sense of *responsibility* for all beings. For Heidegger, ethics is handed down to us based on the Greek idea of a norm, or *ethos*, which *one* is supposed to follow. As submission to a general and culturally evolved norm, ethics may in fact pose a great danger in Heidegger's thinking. But this does not mean that human particularity and integrity are not important to him; on the contrary. As we shall see, it is exactly for this reason that Heidegger does not use a traditional ethical and humanistic framework to determine his view of a responsible human being.

Finally, Latour's and Heidegger's research methodologies are a recurring topic in this book. Since Latour and Heidegger utilize a number of different strategies and approaches in their respective investigations, it is not possible to assess them all within the present framework. The claims being made in this book are that both Heidegger and Latour find inspiration in phenomenology and that their respective research is paradoxically consonant with a certain mode thereof. Phenomenology indeed supports the common point of departure of their work, namely, the fundamental embeddedness of humans in their surroundings, which was mentioned above. Consequently, both ascribe a crucial role to human experience in their writing. Phenomenology prioritizes lived experiences and careful descriptions of world encounters that are neither subjective nor objective, but highlight what Latour calls the "kingdom in the middle" (Latour 1993, 78 and 89; Blok and Jensen 2012, viii) and are addressed in classic phenomenology as "lifeworld." Though Latour distances himself in words from the concept of phenomenology, it will be argued that the scope and completion of his actual work incorporate fundamental phenomenological characteristics. It is also noteworthy that Heidegger's early thinking is explicitly more phenomenological than his later work, which is characterized by its long, winding, and meticulous etymological disentanglements and reflections on history. Yet, Heidegger also viewed etymological inquiries as an attempt to associate a concept with its lived meaning in different historical sceneries. Testifying to this fact, not least, is his famous analysis of the Greek origin of the concept of truth as revealing, *alêtheia*, which gives back a clear and retractable lifeworld signification to the concept of truth—an issue that will be addressed in further detail and from various perspectives in different parts of the book.

The book is divided into six chapters, and it should be approached in one of two ways: Either read in the order presented or the reader should begin with chapter 5, which explains the valuable notion of phenomenology based on some of Latour's more recent work. This alternative way of reading would clarify some of the key methodological and philosophical vocabulary for readers less familiar with Latour and Heidegger.

The first chapter, "Monsters of Modernity," presents my original interest in comparing the writings of Heidegger and Latour.[4] A number of the topics presented here are developed later in more depth, altered, or otherwise reassessed in subsequent chapters on Latour and Heidegger. Most importantly, this chapter argues that Heidegger's notion of enframing and the dangers he associates with modern technology may actually be located in Latour's own version of technical mediation, contrary to Latour's own claims. In a certain sense, I show how Latour falls prey to his own allegations against Heidegger. In Latour's work, it is thus possible to identify and further develop a rearticulation of the sort of danger associated with modern enframing, which Heidegger originally uncovered and warned against.

The subsequent chapter is entitled "Modernity in Action." One of Latour's most renowned books is entitled *We Have Never Been Modern*. Contrary to this claim, Heidegger stresses the threat of modern technology, which arguably confronts us today more than ever. In this chapter, I develop an overlooked relation between Latour's concept of a network and Heidegger's notion of an artwork, and use this to substantiate my claim that the two thinkers are, in fact, trying to overcome the pitfalls of modernity in a very similar way. By drawing on Heidegger's understanding of the work of art, we can make new associations between Heidegger and Latour. This novel interpretation of Heidegger's understanding of modernity and its consequences has escaped the eyes of Latour.

In the third chapter, "Death and the End of Networks," Heidegger and Latour are compared in philosophical-anthropological terms. Latour's work has been linked to the concept of networks and the transgression of seemingly strong divides. In this chapter, I present and critique the Latourian concept of humans; his alleged symmetrical anthropology is analyzed in light of the challenge posed to it by Heidegger's concept of death. The key question is whether Latour's network can overcome the *great divide* between life and death and, if so, what influence this may have on our understanding of human life and what it means to be human. I argue that Latour's framework of human life cannot account for the radical difference between life and death proposed by Heidegger, and that it never intended to do so. In this light, Latour's humans seem to turn into more or less well-functioning cybernetic puppets, who are easily replaceable. In this modern drama, individuality and authenticity become outdated anthropological notions.[5]

In the fourth chapter, I reconcile some of the fundamental oppositions between Heidegger and Latour by suggesting a new starting point for comparison. For this reason, the chapter carries the title "Another Beginning." The apparent opposition between Latour and Heidegger may largely be due to the fact that Latour primarily refers to Heidegger as the author of "The Question Concerning Technology." If we take a different point of departure and focus on Heidegger's renowned lecture "The Thing," where his phenomenological thoughts are presented in a thorough manner, we may arrive at a different understanding and learn to see how Heidegger unfolds the existential meaning of things from a phenomenological perspective, which may be able to save us from some of the fundamental dangers of modernity. This version of Heidegger's thinking proves to be more compatible with Latour's positive notion of religion. Following this line of thinking, it is also shown how Heidegger's concepts of "the fourfold" and "poetic reasoning" may be regarded as a common approach to thinking for both Latour and Heidegger, which elicits mutual associations instead of standing in their way.

The fifth chapter addresses the methodological considerations of Latour and Heidegger running more or less explicitly throughout the entire book. It focuses on what may arguably be called Heidegger's most important early work and Latour's most important late work, namely, *Being and Time* (Heidegger 1927) and *An Inquiry into Modes of Existence* (Latour 2013). I maintain that these works are not only equal in their philosophical ambition, but that they build on and extend phenomenology as a sound and successful approach to philosophical investigations. Although this claim is contrary to Latour's proclaimed self-understanding, I argue that especially *An Inquiry into Modes of Existence* offers many reference points to sustain such a claim, and that this book presents an excellent opportunity to reassess Latour's oeuvre and to reread him as one of the most original heirs of Heidegger's thinking.[6]

The sixth and final chapter of the book considers Latour's most recent interests and shows how Latour, based on the unfolding environmental disaster of the world, ends up articulating a critique of modernity with the same gravity as Heidegger's critique. Latour awakens from his nonmodern slumber, looks at the current state of affairs in the world, and explicitly challenges the modernizers in a fight that is equal in significance to the one launched by Heidegger. By the end of the book, I hope to have shown in detail two strong testimonials from two of the most eminent thinkers of the twentieth and twenty-first centuries, warning us against the profound dangers of modernity. However, both advise their readers and arm them with weapons for counter-construction.

The book seeks to articulate a philosophically significant Heideggerian response to Latour in a way that values and takes seriously Latour's own insights. For this reason, it might seem as if I am siding with Heidegger at times, though my main objective is to better understand the fundamental topics surrounding modernity.[7] It was important for me to compose a Heideggerian response to Latour because it is indeed possible to present a strong and timely reply based on his thinking and because he died before he had the chance to read and comment on Latour's work. At the same time, this approach also means that I take Latour seriously as a prominent philosopher and believe it is important to discuss his works in relation to Heidegger's.

Even though Latour was getting older and had cancer while I was writing this book, I was not anticipating his death; ironically, the last chapters highlight the change that took place in Latour's later writings, which brings both of their critiques of modernity even closer and allows the two giants to be at peace with one another. The contents of the chapters vary, but all reflect a number of significant concerns about modernity. It is important to note that while each chapter may be read on its own terms, it is only in the framework of the entire book that the multiple nuances and different layers of this comparative work on modernity take shape.

In this introduction, some methodological considerations underpinning the philosophical inquiries of the book are elucidated, as well as further reflections on what could be termed "theoretical fieldwork" (Riis 2008). Theoretical fieldwork should not be construed as an attempt to undermine the empirical turn or advocate for the cessation of empirical research in philosophy in general or the philosophy of technology in particular. However, it is pertinent to note that research does not need to be empirical in a traditional sense to be deemed valuable and capable of generating novel insights. It is furthermore essential to underscore the significant body of valuable empirical research that has already been conducted. Equally important is the necessity to pause, reflect upon the amassed material, and endeavor to synthesize it in various ways. The *field* of theoretical fieldwork comprises thoughts materialized in more or less tangible forms of expression, and the study, interpretation, and experimentation with these forms of expression constitute a laborious endeavor bearing many resemblances to empirical fieldwork. Moreover, if we aspire to stand on the shoulders of past giants in order to assess and engage with their ideas, then we must conduct a form of theoretical fieldwork.

To execute the envisaged comparative philosophical fieldwork, it has been imperative to explore and comprehend the specific framework of Latour's and Heidegger's governing concepts and principles as they are elucidated in the selected texts. Philosophy and philosophical thinking are primarily manifest in written language, thus being inherently textually mediated. The articulation of principles, the explication of experiences, and the construction and elaboration of arguments constitute pivotal components of philosophical praxis.

Because fieldwork is conventionally associated with the empirical sciences, particularly anthropology, ethnography, and science and technology studies, and some scholars regard it as a defining criterion of these disciplines, the concept of theoretical fieldwork may initially appear paradoxical; however, viewing a text as a vast yet specific type of field and drawing upon insights from conventional fieldwork in its analysis is very generative. Therefore, given Latour's extensive engagement in fieldwork and his high regard for it, I propose to coin and conceptualize my methodology as theoretical fieldwork, despite its resemblance to other hermeneutic and analytical approaches. The notion of theoretical fieldwork offers several significant and closely interconnected advantages: (1) It establishes a common foundation for both theoretically and practically oriented research from the outset, facilitating the attainment of a shared vantage point for scholars from diverse research domains, such as Latour and Heidegger. (2) It renders highly theoretical texts more accessible, slows down the reading process, and counters what, inspired by Latour, could be termed "the double-click fallacy of reading," which perceives language merely as immediate information without transformation (cf. Latour 2013). (3) Theoretical fieldwork underscores the importance of probing, experimenting, and writing during the research process. This method may also imply extracting longer quotes in order to scrutinize particular phrasings and thus evidence the voice of the indigenous, that is, the original thinker in question. Theoretical fieldwork also enables reading an author against their own professed self-understanding, as is the case in the present investigation concerning Latour and Heidegger. (4) It fosters an appreciation of the practical tools and techniques employed in traditional theoretical work, which regards words and concepts as specific instruments of thought. Notably, Latour expressed, "The text, in our discipline, is not a story, not a nice story, it's the functional equivalent of a laboratory" (Latour 2004b, 69). (5) Finally, it facilitates the identification of guiding principles and assumptions underlying various forms of academic and empirical work, uncovering inconsistencies and latent controversies within a single text.

While theoretical fieldwork shares commonalities with classic hermeneutics, exegesis, and discourse analysis, it pivots on the practical aspects of theoretical work more prominently than the latter three approaches. Moreover, it minimizes the distinction between empirical and theoretical sciences from the outset, suggesting a symmetry between theoretical and practical endeavors—a perspective particularly advantageous for research endeavors such as those undertaken in this book. Classic fieldwork, characterized by extensive immersion and observation in the field, originally conducted by Franz Boas and Bronisław Malinowski to extricate scholars from the confines of books and familiar academic institutions of the early twentieth century, readily translates to classic hermeneutics, mandating thorough and repetitive readings, akin to what Nietzsche, in field-metaphor parlance, refers to as "rumination" in the preamble to *The Genealogy of Morals*, "One skill is needed—lost today, unfortunately—for the practice of reading as an art: the skill to ruminate, which cows possess but modern man lacks. This is why my writings will, for some time yet, remain difficult to digest" (Nietzsche 1956, 157).

A central objective of classic fieldwork in anthropology is to unveil and elucidate the insider's perspective, taking into consideration the context and intricacies of the lifeworld of those under investigation. Similarly, in classic hermeneutics and what I term theoretical fieldwork, this necessitates approaching a text with utmost impartiality or self-reflexivity, conducting an immanent interpretation that encompasses the field of observation and interaction to the greatest extent possible. In theoretical fieldwork, this primarily entails an intrinsic reading in which textual nuances such as emphasis, hesitations, marginal comments, structure, disruptions of structure, uncertainties, paradigmatic examples, references to other authors, and etymology assume critical importance. Simultaneously, this approach does not discount biographical elements as irrelevant, as they may aid in elucidating the intricate context of a text. Biographical elements often pose challenges, as the insider's perspective is frequently ambiguous, sometimes even self-contradictory.

The way I have worked with the texts of Heidegger and Latour in the present book may diverge at some points and seem to go against what I have unfolded as crucial elements of theoretical fieldwork above. This discrepancy may be largely attributed to three related issues: (1) The book presents finished work, which was written over a longer time period and is a product of reiterated readings and interpretations. What seems biased at first sight in the different chapters is in fact in line with the considerations above: The chapters articulate a critical interpretation that slowly developed from rather unbiased readings of Latour and Heidegger. However, there is no purely unbiased reading, and as interpreters, we always carry with us our own rich and complex history. (2) The book also manifests what I call theoretical fieldwork in another sense. It consists of a series of different chapters that *probe* the claims and hypotheses in the shape of writings belonging to theoretical fieldwork, which I have described above. The chapters let me experiment and test my interpretation in a number of different settings, which add layers and robustness to the explanatory framework.[8] (3) Finally, the chapters are accompanied by this introduction in order to make more explicit some of the significant methodological issues shaping the book, that is, render overt what is implicit in each of the individual chapters. In other words, with the notion of "theoretical fieldwork," I have tried to find a methodological approach that would allow for and catalyze the kind of comparative philosophy proposed in this book.

Concluding the introduction to this book, I would like to express my gratitude to the many different persons who have contributed to its fruition. To the researchers, students, friends, and family who have supported me throughout this research journey, I extend my heartfelt thanks. Special appreciation is owed to Peter-Paul Verbeek for his generous feedback on early versions of the manuscript and for graciously agreeing to write the foreword. I am indebted to Graham Harman for sharing his profound insights on Heidegger and Latour, both through writings but also through engaging discussions. I also wish to express my gratitude to the esteemed thinkers whose work and lectures have inspired me over the years, particularly Don Ihde, Günter Figal, and Bruno Latour—whose recent passing is deeply felt. Lastly, my sincere appreciation goes to Rebecca Walsh, whose meticulous editing once again enhanced my work. Her exceptional proficiency as a proofreader has been most gratifying.

NOTES

1. I first read this quote in Graham Harman's original and well-written introduction to Bruno Latour, entitled "Prince of Networks: Bruno Latour and Metaphysics" (2009). Harman writes, "The absence of rats, lions and lakes from mainstream philosophical debate speaks not against Bruno Latour, but against the bland default metaphysics that reduces objects to our human access to them" (Harman 2009, 25). My intention with this reference is to read it autobiographically and somewhat ironically and thus use it to emphasize the difficulty *in knowing* who Bruno Latour really is and my own efforts associating Latour with Martin Heidegger. Interpreted this way, the quote also echoes a later statement of Latour, "'we' no longer know who we are, nor of course where we are, we who had believed we were modern . . . End of modernization. End of story. Time to start over" (Latour 2013, 10).
2. Cf. Kochan (2010).
3. The interesting full quote is as follows:

I was born a Sloterdijkian. When, thirty years ago, I was preparing the proofs of laboratory Life, I had included in the pictures, to the disgust of my scientist informants, a black-and-white photograph of the air-conditioned machinery of the Salk Institute in which I had done my fieldwork. "What does this have to do with our science?" they asked, to which I could only reply: "Everything." Without knowing it, I had always been a "spherologist," as I discovered about twenty years later when I became familiar with Peter Sloterdijk's work in another locally situated, airconditioned place: his school in Karlsruhe, which was separated by no more than one courtyard from the Center for Art and Media, where I twice had the great chance to experiment with installations and exhibitions—what, with Peter Weibel, we call a Gedanke Ausstellung or "thought exhibition," the equivalent in art of a "thought experiment" in science. (Latour 2009, 139)

4. An early version of this chapter has been published in *Social Studies of Science* (Riis 2008a).
5. Together with a former student, Andreas M. Gregersen, I have written an article about what could be called Latour's positive notion of religion (Gregersen and Riis 2015).
6. In this context, postphenomenology, as developed by Don Ihde, Peter-Paul Verbeek, Robert Rosenberger, and others, may also be read as a mediator between

Latour and Heidegger, and Ihde as a progressive heir of Heidegger's philosophy (Cf. Ihde 1995, 2003; Ihde and Selinger 2003; Verbeek and Rosenberger 2015; Verbeek 2016).

7. My book on Heidegger, *Unframing Heidegger's Understanding of Technology*, may indeed be read as a thorough critique of him (Riis 2018).

8. Cf. Latour: "The text, in our discipline, is not a story, not a nice story, it's the functional equivalent of a laboratory. It's a place of trials and experiments" (Latour 2004b, 69).

Chapter 1

Monsters of Modernity

In his renowned book *Pandora's Hope,* Bruno Latour delivers a fundamental attack against Heidegger and accuses him of misunderstanding modern technology and creating a monster out of it, which has taken control of humans: "[According to Heidegger] technology is unique, insuperable, omnipresent, superior, a monster born in our midst which has already devoured its unwitting midwives. But Heidegger is mistaken" (Latour 1999, 176).

To support his claim, Latour's strategy is to show that it is incorrect and even meaningless to speak of the relation between humans and nonhumans—including technical artifacts—as involving any kind of mastery of one over the other. In other words, Latour argues that it is not only deceptive but also wrong to say that technology dominates humans or that humans dominate technology. He demonstrates this claim by playfully reexamining the familiar U.S. National Rifle Association slogan on the relation between humans and firearms, "Guns don't kill people, people do" (Latour 1999, 176). He outlines the meaning of "technical mediation" by constructing four categories, which Latour also thinks show why Heidegger is wrong. To understand Latour's argument, and to begin to see how Latour and Heidegger in fact argue along the same lines but draw different conclusions, we first have to gain an understanding of Latour's four categories of technical mediation. Instead of stressing differences in style and purpose between Latour and Heidegger, which are immediately evident to most readers, I will first seek to reveal some fundamental similarities between the two.

Latour begins his argument against Heidegger by reflecting on two seemingly different understandings of the relation between a person and a gun. While a gun may seem to be nothing more than an instrument that is completely subordinate to the will of the user, Latour argues that it can also transform a neutral bearer into a perpetrator. Admitting that his example is a bit of a caricature, Latour uses it to illustrate how both understandings lead to the same conclusion. The free citizen armed with a weapon is no longer the same as an unarmed citizen, but the weapon in the hand of the citizen is also no longer the same object either—for example, when compared with a gun on a shelf (Latour 1999, 176f). With his concept of "symmetry," Latour wants to clarify and relate the two interpretations: "The myth of the Neutral Tool under complete human control and the myth of the Autonomous Destiny that no human can master are symmetrical" (Latour 1999, 178).

According to Latour, the notion of symmetry means that we are supposed to look beyond both interpretations and pay attention to a third agent who has entered the scene and left the other two behind: the "citizen-weapon" or the "weapon-citizen" (Latour 1999, 179). By speaking of the symmetry between these two novel hybrids or monsters, Latour implies that it neither makes sense to assert that technology subordinates the will of humans nor to claim the opposite. With this analysis, Latour also describes his first category belonging to technical mediation, namely, "interference," and, at the same time, he takes this to be a first step in trying to repudiate Heidegger's concept of technology (Latour 1999, 176f). My claim is, however, the opposite: In doing so, Latour actually substantiates Heidegger's critique of technology.

Anticipating a critique of the first meaning of technical mediation, Latour introduces a second category to his theory of technical mediation, "combination." Latour uses combination to respond to the possible objection that there is a basic asymmetry within the field of interference outlined above: although humans and nonhumans are inseparable, humans are the source of action in relation to nonhumans (Latour 1999, 180). He claims this objection is shortsighted and incorrect. He stresses that production processes, such as the production of computer chips by workers at a factory, involve a number of human actions and tools in different combinations.

A single worker does not make a computer chip. In a long and complicated labor process, individuals combine different tools and link up with the work of others and machines. Generalizing this idea to other complex technologies, Latour claims that certain expressions are misleading, such as "man flies." It is the "man-airplane" that flies. This leads Latour back to his principle of symmetry, which states that it is impossible to separate human actions from nonhuman actions. "Action is simply not a property of humans *but of an association of actants,* and this is the second meaning of technical mediation" (Latour 1999, 182).

To clarify further the meaning of technical mediation, and to reveal the universality of the concept of symmetry concerning technical mediation, Latour goes on to illustrate the third and fourth meanings of technical mediation. Its third meaning is the "folding of time and space" (Latour 1999, 183). With this definition of technical mediation, he points out something quite similar to its second meaning, combination. If a technical artifact consists of and is combined with other technologies, which are themselves combinations of various humans and nonhumans at different points in time and space, then technical devices are to be understood as black boxes (Latour 1999, 183f). Black boxes fold together entities of different times and locations, and each entity also has its own enfolded story. These boxes become even blacker as more humans and nonhumans are "bent, enrolled, enlisted, mobilized and folded" into them and then displaced further away in time and space (Latour 1999, 185f).

Latour considers the fourth definition of technical mediation as the most important one because it subsumes the other three (Latour 1999, 185). This mode of technical mediation consists in *crossing the line between signs and things*, and he claims that technological devices can express normativity in a way that differs from traditional written or oral discourse (Latour 1999, 185). Through such technical articulation, it becomes possible to bridge the gap between things and signs, as well as the gap between *is* and *ought* and thus to introduce normative affordances into the fabric of the world.

Engineers are the true masters of such technical articulation: they develop techniques to meld things, demands, and meanings. Latour illustrates this fourth meaning of technical mediation with a speed bump—a "sleeping police officer" or hump of pavement designed to slow down the speed of cars (Latour 1999, 186). The language of the speed bump is unmistakable and concrete, that is, it is a special kind of technical mediation that delegates meaning. Instead of hiring police officers to control speed all day and night, traffic engineers have delegated this work to speed bumps. The agency of the speed bump can be understood as a complex of a number of humans and nonhumans, ranging from the police officers, engineers, politicians, and construction workers to the number of different materials taken from various places. The speed bump is a certain kind of black box or technical delegate insofar as it redistributes the absence and presence of a number of various actants and interferes directly with the daily life of urban car-humans (Latour 1999, 189).

Latour's four categories of technical mediation shape his critique of Heidegger's conception of technology. In summary, human and nonhuman actions are impossible to separate and mutually dependent; furthermore, such a separation is not even desirable. What really fascinates Latour is the capacity to make new black boxes that recombine human and nonhuman agency in order to solve practical problems. In other words, he ultimately tries to facilitate what could be called a *political ontology* or a program of *planet management*. According to Latour's account of the increasing entanglement of humans and nonhumans through technical mediation, "*Literally*, not symbolically as before, we have to manage the planet we inhabit" (Latour 1999, 203). Not surprisingly, Latour's protagonist is the ancient prototype of the engineer, Daedalus: the cunning trickster who managed to navigate his way through a previously inescapable labyrinth, constructed military robots, and created the technology to kill King Minos, who had imprisoned him (Latour 1999, 175f).

Playing with concepts from traditional philosophical discourse, Latour argues

> Objects and subjects are made simultaneously, and an increased number of subjects is directly related to the number of objects stirred—brewed—into the *collective*. The adjective modern does not describe an *increased distance* between society and technology or their alienation, but a deepened *intimacy*, a more intricate mesh, between the two. (Latour 1999, 196; emphasis added on "collective")

Latour's critique of Heidegger's notion of technology in the initial quote thus culminates with the idea of the ultimate symmetry of *the collective*—a major black box—inhabited by metamorphosed and inseparable human and nonhuman agents who have been mobilized, brewed, and folded irreversibly together in time and space. We may see this collective as the production site of hybrid creatures of all flavors—of monsters of modernity.

TECHNOLOGY AT WORK

Martin Heidegger showed remarkable interest in technology over the course of his life, which appeared in his very first books and lectures as well as his last (cf. Heidegger 1969, 2002a). As presented above, Latour frames his critique of Heidegger with four concepts of technical mediation, the most important of which is "crossing the line between signs and things," that is, eliminating the boundary between behavior and artifacts (Latour 1999, 185). Heidegger's investigation of technology "at work" centers on a single concept, namely, his famous concept of enframing (*Gestell* in German). But he also produces four key notions for understanding technological mediation (Heidegger 1977a). With the concept of enframing, Heidegger describes what he views as the most fundamental characteristic of technology and technological mediation: the essence of technology. He is not interested in what is distinctive about one or another type of technology; instead, like Latour, he is concerned with the common trait—the essence—of technical artifacts, technology at work, or technical mediations. To study the essence of technology, Heidegger finds one must avoid becoming preoccupied with any single technical artifact. In empirical studies, the concreteness of case studies can easily be blinding, so that it becomes difficult to see the bigger picture.[1] Just as Latour's concept of "technical mediation" highlights what belongs to all technologies in terms of a larger framework of technical processes, so does Heidegger's concept of enframing. The essence of technical devices is neither the device itself nor a substance of any kind. The essence of technology is rather to be understood as a process, or as a certain way of bringing something into appearance, which can be analyzed in terms of means and ends (Heidegger 1977a, 311).

In order to better understand Heidegger's investigation of the essence of technical mediation, and to uncover the profound similarities it shares with (and even anticipates) Latour's research, we shall now take a closer look at Heidegger's key notions. The first notion concerning the essence of technology or technology at work is "revealing". Technology is, first and foremost, a way of bringing something into appearance, "What has the essence of technology to do with revealing? The answer: everything. For every bringing-forth is grounded in revealing" (Heidegger 1977a, 318).

Just as every kind of revealing or bringing-forth carries a special signature, so too is revealing mediated through technology. What is brought forth through technology receives a specific identity. Heidegger wants to investigate the specific character of this revealing process, that is, technical mediation, alongside three other basic concepts.

His investigation sets out at a very fundamental level. To be able to understand why Heidegger proceeds this way and places the concept of revealing at the beginning of his research on technology, we will briefly dwell on another concept more familiar to science and technology studies: the Kuhnian concept of "paradigm." This should help clarify Heidegger's fundamental concept of revealing. A paradigm *governs* how phenomena are conceptualized and investigated in a given scientific field at a certain time. A shift of paradigm revolutionizes the understanding of the world, as it *reveals* the world differently, "In so far as their [the scientists'] only recourse to that world is through what they see and do, we may want to say that after a revolution scientists are responding to a different world" (Kuhn 1970, 111). Translated into Kuhnian terms, Heidegger's investigation proposes a specific paradigm of technology, its *essence*, which he subsequently attempts to describe. Heidegger claims that the essence of technology is a process that frames the way we interact, think about, and visualize the world.[2] In other words, to support this claim about enframing, Heidegger develops three crucial concepts. Explicating these concepts should help us see how Latour's account of technical mediation is in agreement with Heidegger's understanding of technology at work. Heidegger elaborates the *revealing* of technology using the notions challenge, standing-reserve, and ordering (Heidegger 1977a, 311). "Challenge", for Heidegger, means that demands are made on the world as it is revealed through technology. When revealed through technology, nature is understood and visualized as a resource, "The revealing that rules in modern technology is a challenging, which puts to nature the unreasonable demand that it supply energy which can be extracted and stored as such" (Heidegger 1977a, 320).

To avoid any misunderstandings of what Heidegger means by challenge, it is important to understand that energy is not only to be understood in the sense of modern physics, but also as a description of any flexible resource that can be instrumentalized further and transported to other locations and kept in storage for later use. This notion of a challenging-revealing of nature as an energy resource also helps us to understand what Heidegger means by "standing-reserve". Through challenging-revealing, nature appears first and foremost as a resource that is made to be used, changed, and answer to the demands made upon it. When Heidegger describes the world revealed through technology as a standing-reserve, the concept of standing should not be associated with something inflexible or static. The world revealed as standing-reserve is not at all static; it is mobilized and ready; it is prepared and now *stands* under command. In this way, energy as unfolded above belongs to what Heidegger views as standing-reserve.

The concept of standing-reserve is also connected to Heidegger's third concept of revealing, "ordering." The standing-reserve is ordered to make even more resources available, and in this sense, the circle of revealing and utilizing resources is completed. The ordering of new resources—human or nonhuman—is furthermore catalyzed through a calculative and cybernetic way of thinking (Heidegger 1999b, 107f). When compatible and reduced to a common denominator, the resources are easier to control, and the process of ordering is optimized and may become more efficient. Heidegger's famous description of the Rhine is significant in light of his concept of "revealing through technology," as well as the three concepts—challenge, standing-reserve, and ordering—which he uses to specify its meaning.

The hydroelectric plant is set into the current of the Rhine. It sets the Rhine to supplying its hydraulic pressure, which then sets the turbines turning. This turning sets those machines in motion whose thrust sets going the electric current for which the long-distance power station and its network of cables are set up to dispatch electricity. In the context of the interlocking processes pertaining to the orderly disposition of electrical energy, even the Rhine itself appears to be something at our command (Heidegger 1977a, 321).

Following Heidegger's argument, the hydroelectric plant reveals the Rhine by challenging its current to deliver energy, which in the end can be used to bring even more resources under control. This example of technical mediation is paradigmatic in Heidegger's text. Since the Rhine cannot immediately and without difficulty be transformed into a source of hydroelectric energy, it might take a long time to achieve control of its resources, but this is only considered a preliminary *problem* that will eventually be *solved* (Heidegger 1999, 76). If this self-supporting circuit of preparing and using natural resources continues, the boundaries between technology and nature ultimately are undermined, overridden, and torn down. The goal of the process initiated by enframing is to mobilize every-*thing* to support this circuit and thus to create a complete symmetry between nature and technology. Based on this example, the human workers constructing the power plant place demands on the Rhine; for Heidegger, equally important is that when doing so they are themselves challenged by the essence of technology to improve the technical efficiency of the hydroelectric power plant. The construction workers are very similar to Latour's factory workers, who coproduce and combine different technologies into computer chips. Even though the human factory workers immediately appear to be the real actants of the construction process of computer chips, they are actually just parts of a larger process consisting of materials and technologies, which governs their behavior most of the time. In these overarching processes, there are numerous active players or actants, such as the Rhine, the computer business, and the construction equipment, which all play their specific roles. Revealed by enframing, Heidegger would agree that humans and nonhumans indeed become symmetrical in Latour's sense. Stated differently, when enframing has been doing its job, symmetry between humans and nonhumans prevails.

HUMAN RESOURCES

We have now gone through some of Latour's and Heidegger's crucial arguments, concepts, and examples concerning technology and technical mediation. To unearth the similarities and concurrences, so that the relationship between their descriptions can be grasped, we must (1) look beyond their different terminologies along the lines of the preliminary reflections on theoretical fieldwork and (2) identify the logic or way of thinking of their parallel conceptions of technology. To begin to address the relationship, I will examine Latour's and Heidegger's arguments from two related points of view. First, I will compare a way of thinking that, according to Latour, is capable of reaching the goal of what can be called the post-human collective with Heidegger's understanding of the challenging and ordering of the essence of technology. Second, I will compare Latour's ontology of the collective with Heidegger's writings on the world as revealed and ruled by enframing.

Latour describes his protagonist Daedalus as being able to find "solutions where none are visible, using any expedient at hand" (Latour 1999, 190). He is able to turn anything into a tool and to combine tools into complex apparatuses. Playing cunning tricks, he turns an ant, some honey, and a mere thread into a navigational tool. Later on, according to the ancient myth, Daedalus transforms what is supposed to be a pleasant shower for King Minos into a cruel killing device. In line with these remarkable and powerful skills, Latour's Daedalus is also portrayed as the great-grandfather of genetic engineering—a pioneer of monsters—enabling "Poseidon's bull to impregnate Pasiphae to conceive the Minotaur" (Latour 1999, 175). At his will, Daedalus changes and instrumentalizes his surroundings, and finds solutions where none are visible to other people. Not only does he *seek* solutions, but he is also perpetually engaged in *constructing* them: his constructive approach becomes a fundamental way of reasoning and revealing and manipulating beings. Accordingly, Daedalus sees the world in terms of challenges and problems, on the one hand, and solutions and constructions, on the other.

It is the descendants of Daedalus who are now developing techniques to petrify police officers into stone-hard speed bumps or to genetically CRISPR-engineer new modern monsters. In the hands of these contemporary polytechnicians, the world has become plastic and manipulable at will. To catalyze their capacities, Daedalus' epigones have developed various calculative devices, such as computers and AI programs. Following Latour's Ariadne's thread, neither humans nor computers think on their own; rather, the *computer-humans* calculate and the large human language models write. And if these hybrid humans who make up our institutions decided not to set up speed bumps and to rely on human police officers instead, they would still educate, regularize, and standardize them to stand at attention when ordered. Humans appear as cyborgs, that is, as complex, manipulable organisms. The difference between speed bumps and police officers is important, but the general desire to meld them in order to control and regulate society must not be overlooked. With this clarification of Latour's notion of technical mediation, we have at the same time reached a description very similar to Heidegger's notion of the challenging-revealing of the world and its enforcement through ordering.

For Latour, the ultimate result of technical mediation is the formation of a grand collective—and such a collective does not exclude the sort of dystopian view of modernity that breeds humans and nonhumans into a giant stockpile of all sorts of resources. The ability to frame the world according to means and ends and to reason in terms of problems and solutions, as Latour's protagonist Daedalus epitomizes, supports this mobilization process and belongs to what Heidegger captures with the concept of ordering according to enframing.

In a visionary statement, Heidegger questions the way humans are challenged and absorbed by the essence of technology:

> Only to the extent that man for his part is already challenged to exploit the energies of nature can this revealing that orders happen. If man is challenged, ordered, to do this, then does not man himself belong even more originally than nature within the standing-reserve? The current talk about *human resources*,

about the supply of patients for clinics, gives evidence of this. (Heidegger 1977a, 323; emphasis added)

Heidegger epitomizes this challenging-revealing of humans with a quotation from Wendell Stanley, who won the 1946 Nobel Prize in Chemistry for his research in biochemistry on the structure of viruses: "The hour is near, in which life is laid in the hands of the chemist who can decompose, build and change this living substance at his wish" (quoted in Heidegger 1966, 20). The human engineer or the descendant of Daedalus is capable of seeing and interacting with basic elements of life as a resource—a conception that can extend to human life and open previously unthinkable possibilities of genetic engineering.[3] For example, how might a Minotaur of the twenty-first century look?

Following Heidegger's argument concerning the rule of enframing, there is neither an *essential* difference between humans and artifacts nor a concept of freedom that runs through the endless chain of orders and orderings. Ethics appears in the shape of *values*, which only confirm the ability to evaluate, calculate, recombine, and substitute all things, especially the things we care about, to the benefit of the effective production and consumption of resources. As long as the total value remains the same or more, there are no real ethical problems, on the contrary (Heidegger 1977a, 249). Considering the ultimate form of technical mediation, Latour and Heidegger agree that any concept of objects or subjects loses its meaning and becomes obsolete. For Latour, the collective folds together humans and nonhumans. For Heidegger, enframing challenges and governs beings of all kinds and does not distinguish between humans and nonhumans in terms of the standing-reserve that exhausts all of its properties when preparing them to be ordered as resources.[4] In this sense, humans and nonhumans become impossible to thoroughly distinguish. With an example that remains familiar in the twenty-first century, Heidegger explains how the concept of a nonhuman object—here meant as an independent thing with integrity, standing over and against a human being—in fact loses its independence and thus its explanatory power due to the essence of technology:

> Yet an airliner that stands on the runway is surely an object. Certainly. We can represent the machine so. But then it conceals itself as to what and how it is. *Revealed*, it stands on the taxi strip only as *standing-reserve*, inasmuch as it is *ordered* to insure the possibility of transportation. For this it must be in its whole structure and in every one of its constituent parts itself on *call for duty*, i.e., ready to take off. (Heidegger 1977a, 322; emphasis added)

An airplane makes no sense in itself but only within the framework of transport and the management of resources standing ready for use. Within the standing-reserve of resources, humans and airplanes are melded and produce each other.[5] Thus, for Heidegger and Latour, it only makes sense to say that it is the human-airplane that flies.

The essence of technology reveals a world of interconnected beings in a pervasive large-scale structure: nothing stands for itself and is able to resist the challenge of being ordered as a means in the process of managing and optimizing the readiness of resources in the world. Everything turns into a kind of tool. This is Heidegger's idea of a technically mediated collective. He views this as the supreme danger because it reduces human beings to mere cogs in a machine and hence prevents them from realizing the highest potential of their being, namely, as guardians of an understanding of truth as something that is procedural and essentially open to change (Heidegger 1977a, 333). The working of modern technology, or enframing, instead seeks to enslave humans and nonhumans alike under one and the same one-dimensional revealing of being—a revealing that turns the Rhine into a power source and humans into computers and speed bumps.

From an ontological point of view, the monsters and hybrid beings who inhabit Latour's collective have the same character as Heidegger's beings in the chain of command governed by enframing.[6] The Daedalus of our day interprets all beings as temporary obstacles or as a means to solve problems—*being* is increasingly mobilized. This mobilization process is enforced through what Latour describes as "delegation," "The act of transportation has been shifted down, not out—down to planes, engineers and automatic pilots, object-institutions to which has been delegated the task of moving while the engineers and managers are absent (or limited to monitoring)" (Latour 1999, 188f). If modern air transportation is to work properly, thousands of actants have to be coordinated, controlled, and pushed in the same direction. Though these acts of (transport) delegation have increased over the last few decades, they have in principle always taken place, according to Latour, and now form our instruments, intuitions, and traditions. Individual agency, intentionality, and freedom are completely dominated or guided by collective agency, which over centuries has been invested into the collective fabric of the world and now includes compounded interest (Latour 1999, 197). Latour argues:

> *Purposeful action and intentionality may not be properties of objects, but they are not properties of humans either.* Only corporate bodies are able to absorb the proliferation of mediators, to regulate their expression, to distribute skills, to force boxes to blacken and close. Objects that exist simply as objects, detached from collective life, are unknown, buried in the ground. (Latour 1999, 192f; emphasis added)

The collective extends its active networking to everybody and everything.

The main focus of human beings and aim in life becomes to have a good job, that is, to have a clear function within the collective, to solve its (more or less partial) problems, and to reinforce its progression.

> If anything, the *modern* collective is the one in which the relations of humans and non-humans are so intimate, the transactions so many, the mediations so convoluted, *that there is no plausible sense in which artefact, corporate body and subject can be distinguished.* (Latour 1999, 197; emphasis added)

After completing his interpretive work on technical mediation, Latour finds that individuality has vanished and that all humans have been overruled by the collective. Humans are merely defined as a temporary constellation of properties divested of *purposeful action* and *intentionality. All human properties can, in principle be transferred to nonhumans—there is nothing essential to being human.*[7]

Latour's collective is a seamless and pervasive web of humans and nonhumans that are conceptualized as resources; it is a hybrid monster, a sort of Leviathan in French uniform. Latour does not use the concept "resource," but he only focuses on humans and nonhumans based on their function as actants, and their work is rendered intelligible through the network, that is, as a resource for the collective. It is only in this structure, in this network, that humans and nonhumans become meaningful. The opposite of resource, something isolated and self-governing, is nowhere to be found in Latour's work.

Latour's notion of technical mediation and Heidegger's conception of the essence of technology reveal and treat all beings as substitutable with artifacts; they can be ordered and manipulated according to the general scheme of mobilization and the collective control. And this mobilization process proceeds toward the ultimate collective or the complete rule of enframing. Ontologically speaking, as the fundamental articulation of beings, Latour's and Heidegger's positions converge and almost merge. In Latour this is mostly studied in local, microscale processes, but the underlying ontology is mobilized, and for the critical interlocutor, it clearly moves in the direction of Heidegger's concept of enframing. Furthermore, if we recall Latour's appeal at the beginning of this chapter, which we will return to in a somewhat different way in the last chapter, he calls for macro-scale planet management, "Literally, not symbolically as before, we have to manage the planet we inhabit" (Latour 1999, 203). For the network to reach planetary proportions and be able to coordinate actants on a global level in a predefined direction, the ontological groundwork needs to be in place and ready to leverage.

In two respects, though, there are important differences between Latour's and Heidegger's understandings of technical mediation, that is, technology at work. Latour accuses Heidegger of turning technology into an *omnipotent, superior monster*, and he abstains from speaking of technology as having a single essence. He instead reduces the meaning of technical mediation to four categories, as previously mentioned: interference, combination, folding time and space, and crossing the line between signs and things.[8]

For a moment, we shall turn away from Latour's fourfold conception of technical mediation and examine the mechanics involved in his development of the collective. It seems at first that he offers no answer to the question—which he does not even pose—of what drives humans and nonhumans into the collective. Nonetheless, by taking a closer look, we can find interesting elements that form an answer. Like a divine and tricky engineer, Latour builds the necessity of technical mediation into the working of *time* itself. Time is the power that forces technical mediations to produce or reproduce the collective and which makes humans reason like Daedalus. In this sense, the agency of time puts Latour's fourfold account of technical mediation on a track that, in spite of its twists and turns, has a clear destination: the ultimate collective. Recall that Latour states, "Instead of clarifying even further the relations between objectivity and subjectivity, *time* enmeshes, at an even greater level of intimacy and on an even greater scale, humans and nonhumans with each other" (Latour 1999, 200; emphasis added).

Heidegger describes the essence of technology as one way of revealing among others, which we shall study in later chapters. Latour, in contrast, in fact, makes a much stronger claim about the *necessary* connection between technical mediation and temporal transformation. From his argument, it follows that the technical entanglement of humans and nonhumans will become even more intimate: it has become the inescapable destiny of the world. *Time*, ultimately regarded as technical mediation in Latour's framework, is *insuperable, omnipresent, and superior*. All interactions and processes, that is, the working of time, support the development of Latour's collective. However, this expresses a conception similar to Heidegger's own understanding of technology (which Latour initially repudiated), causing him to fall prey to his own allegations. Even Daedalus cannot find a way out of Latour's cunning labyrinth of time.[9] In short, Heidegger's "ordering of enframing" is made absolute by Latour and stated in terms of the agency of time; different terminology is used to describe analogous processes.

The second and more decisive difference between Latour and Heidegger has been evident all along in this investigation, although it has not yet been explicitly mentioned. This difference concerns their judgments on the development of a technically mediated world. Whereas Latour rejoices in technical mediation, Heidegger grieves it and calls it the supreme danger of modernity. Whereas Heidegger is afraid that humans are being reduced to utensils,

Latour welcomes a notion of humans articulated in terms of the properties of tools.

This difference also explains why their styles of writing about technology are so distinct. Latour is full of humor and sprinkles abundant jokes amid his discussions, whereas Heidegger writes in a brooding tone of lurking danger. As we shall see in later chapters, Latour's writing style changes over time and moves in the direction of Heidegger's as he realizes the vast problems for life on earth that Heidegger had already identified.

In the end, Latour's interpretation of technical mediation appears as a sort of mirror image of Heidegger's. In a figurative sense, a symmetrical relation emerges between their respective understandings of a technically mediated world. Their understandings support one another and at the same time oppose each other in a characteristic sense. Both hold the technically mediated world to: (1) fold together humans and nonhumans; (2) challenge them to mobilize the world; and (3) suspend any kind of individual sovereignty. However, this picture is only accurate if we pay close attention to the opposition between Latour's and Heidegger's attitudes—or what readily can be seen as a crucial inversion of judgments—which makes the whole picture all the more thought-provoking. This symmetrical picture of Latour and Heidegger creates a disturbing harmony between two different accounts of technical mediation and points to the idea of the collective *or* enframing as a vital challenge to philosophy and technically imbued contemporary societies. In the next chapter, we shall learn much more about, what this challenge has to do with Latour's and Heidegger's notion of modernity.

NOTES

1. Cf. the presentation of theoretical fieldwork in the Introduction.
2. By being aware of the essence of technology, it is possible to avoid being dominated by it; this awareness enables one to see the similarities between technical *revealing* and other forms of disclosure (e.g., the way of revealing the world, which happens in a work of art). In this sense, Heidegger's investigation of technology, on one level, calls attention to a danger, which we will now look into more closely. But, on another level, Heidegger also addresses a different way of viewing technology. Nonetheless, I will not pay much attention to that level in this chapter, as it does not play an explicit role in Latour's critique of Heidegger; on the contrary, Latour does not acknowledge the fundamental ambivalence toward technology in Heidegger's argument. In Latour's book *We Have Never Been Modern* (1993), he falsely criticizes Heidegger as one of the moderns. He writes, "The moderns indeed declare that technology is nothing but pure instrumental mastery, science pure Enframing and pure Stamping (*Das Ge-stell*), that economics is pure calculation, capitalism pure

reproduction, the subject pure consciousness. Purity everywhere!" (Latour 1993, 66). For Heidegger, the essence of technology discloses a fundamental ambivalence of technology. On the one hand, it is capable of revealing *beings* and creating a *world*— a capacity technology shares with artwork—on the other hand, it reveals beings in a special way, which we examine in this chapter. Furthermore, Heidegger also explicitly rejects that technology is pure instrumental mastery, which Latour accuses him of saying. In Heidegger's conception of technology, there is *ambivalence* in the place of *purity*! (Heidegger 1977a, 313f). Cf. Glazebrook (2000).

3. It is also telling that the Nobel Prize in Chemistry in 2022 was awarded for successfully working with molecules as *Lego building blocks* (Rannard, 2022). Cf. also Anderson et al. (2005). However, the suggested fixation of nature in biotechnology is not more stable than it may be destabilized over time—just as apple trees may not yield any apples some years. Furthermore, new stabilities attained through biotechnology are often more fragile than old ones, which have adapted to their surroundings and proved capable of living over longer periods of time. From this perspective, uncertainty, not stability, may also be seen at the heart of biotechnology and designate new openings. To understand more about the political implications of this uncertainty see Jasanoff (2005, 123f).

4. This is also clear to Don Ihde, who describes the *symmetrist's error* as the temptation *to mechanize* the totality; or *to socialize* it (Ihde, 2003, 143). Harman writes, "Latour is reluctant to believe that anything substantial could exist outside of all networks" (Harman, 2002, 312). Compare with Zammito (2004, 201).

5. In order to see how an airplane can be revealed differently, cf. chapter 4.

6. Latour mentions a real, thought-provoking encounter with what could be called a real part of the collective:

> To illustrate: some time ago, at the Institute Pasteur, a scientist introduced himself, "Hi, I am the coordinator of yeast chromosome 11." The hybrid whose hand I shook was, all at once, a person (he called himself "I"), a corporate body ("the coordinator"), and a natural phenomenon (the genome, the DNA sequence, of yeast). The dualist paradigm will not allow us to understand this hybrid. (Latour 1999, 203)

And if we read further, then we also get a sense of how the scientist and the yeast serve the collective:

> Still, the industrial laboratory where he works is a place in which new modes of organization of labor elicit completely new features in nonhumans. Yeast has been put to work for millennia, of course, for instance in the old brewing industry, but now it works for a network of thirty European laboratories where its genome is mapped, humanized, and socialized, as a code, a book, a program of action, compatible with our ways of coding, counting and reading, retaining none of its material quality, the quality of an outsider. (Latour 1999, 203; emphasis added)

Here we see what it means to be *socialized* by the *collective*. The yeast and the scientist are equally challenged through this *research* in order to *work* as efficiently as possible and serve the collective in the best possible way.

7. Latour says, "In abandoning dualism our intent is not to throw everything into the same pot, to efface the distinct features of the various parts within the collective"

(Latour 1999, 193). Latour's *intentions* are not at stake, but how he literally interprets *symmetry* within the *collective*. What I want to stress is that, for Latour, the *distinct features* of the *various parts* are only temporary and can, in principle, be changed and mobilized at any time. And if the *distinct features of the various parts* are constantly changed, modified, and mobilized within the collective, what is then so distinct about the various parts? For example, at one place Latour argues, "Provisional 'actorial' roles may be attributed to actants only because actants are in the process of exchanging competences, offering one another new possibilities, new goals, new functions" (Latour 1999, 182).

8. I have left aside the fact that Latour treats *technical mediation* as essentially one kind of act, but with four different meanings. He talks about the first, second, third, and fourth meanings of technical mediation and keeps using the term *mediation* in the singular. At the same time, Latour (1999, 185) constructs a hierarchy among the meanings of technical mediation so that the fourth meaning is interpreted as the most important one. For these two reasons, Latour's *fourfold explanation* of technical mediation can be interpreted as, in fact, *onefolded*.

9. Over time, Latour's *collective* makes everything symmetrical, forcing all things to take on a uniform shape from which there is no turning back and no room for significant differences. Latour stresses in this account of technical mediation that "the arrow of time is still there" (Latour 1999, 200). At first glance, this may seem harmless, but it really means that all beings are challenged and manipulated in a similar way, that there is something in control of all beings which has a specific direction and is always active.

Chapter 2

Modernity in Action

Martin Heidegger declared that the essence of modern technology reveals what he understood to be the ultimate danger. A few decades later, at the end of the twentieth century, Bruno Latour proclaimed that "we have never been modern." It seems that these two statements are contradictory, as the former ascribes the highest threat to what the latter rejects as an illusion. Furthermore, Latour explicitly critiques Heidegger and claims he is a key opponent in his programmatic denunciation of modernity. However, as we shall see more clearly in this chapter, both thinkers fundamentally disapprove of whatever they consider *modern*. And both of them also view the concept of modernity as crucial to their own thinking. Finally, both Latour and Heidegger strive to bypass the dead ends associated with this historical era. In other words, readers shall not be led astray by Latour's apparent critique of Heidegger, but be reminded of the sort of theoretical fieldwork encouraged in the introduction of this book. The focus shall be on the course of his arguments, as well as the numerous cues and peculiar concepts apparent in his renowned text on modernity.

In this chapter, I will show what is at stake in this fundamental tension between Heidegger and Latour. At the same time, I endeavor to answer the following two questions: (1) What is the meaning of modernity for Heidegger and Latour? (2) What, if anything, do technologies, artworks, and networks have in common? Sorting out these issues and answering these two questions shall help us to answer the seemingly simple question, *Have we never been modern?*[1] The guiding thesis of this chapter sides with Heidegger insofar as it claims that modernity is no mere illusion and that the powerful technologies associated with it do indeed change how and what we can relate to. . However, the questions guiding this chapter is are tricky one, and Latour is a cunning thinker, so we have to move slowly and pay careful attention to how we articulate an adequate response. By the end of this chapter, my aim is to add additional layers to our preexisting understanding of both Heidegger and Latour.

LATOUR'S NONMODERN CONSTITUTION

Latour's highly influential book *We Have Never Been Modern* (Latour 1993 [1989]) sets out to redefine ontology, the kinds of beings inhabiting the world, and how they interact and connect. Despite its vast impact on sociology and science and technology studies, and its explicit philosophical scope and references, the book has not yet received much attention in philosophy. In this book, Latour launches his critique of the moderns, outlines the foundation of an alternative ontology, and describes the operation of the two crucial processes of what he calls "purification" and "hybridization."

Latour's connection and opposition to Heidegger are significant to how he understands his own project and frames his own position:

> We are carrying out the impossible project undertaken by Heidegger, who believed what the modern Constitution said about itself without understanding that what is at issue there is only half of a larger mechanism which has never abandoned the old anthropological matrix. *No one can forget Being, since there has never been a modern world, or, by the same token, metaphysics.* We have always remained pre-Socratic, pre-Cartesian, pre-Kantian, pre-Nietzschean. No radical revolution can separate us from these pasts, so there is no need for reactionary counter-revolutions to lead us back to what has never been abandoned. (Latour 1993, 67; emphasis added)

Here, Latour clearly states that no specific modern world exists; there has been no radical revolution in human history, according to Latour, and he maintains that Heidegger is misguided in thinking that modernity has a special character.[2] In a further attempt to distance himself from Heidegger, Latour quotes Heraclitus and claims that the *gods* also are present in modern technology:

> And yet—"here too the gods are present": in a hydroelectric plant on the banks of the Rhine, in subatomic particles, in Adidas shoes as well as in the old wooden clogs hollowed out by hand, in agribusiness as well as in timeworn landscapes, in shopkeepers' calculations as well as in Hölderlin's heartrending verse. But why do those philosophers no longer recognize them? Because they believe in what the modern Constitution says about itself! This paradox should no longer astonish us. The moderns indeed declare that technology is nothing

but pure Stamping [Das Ge-Stell], that economics is pure calculation, capitalism pure reproduction, the subject pure consciousness. Purity everywhere! They claim this, but we must be careful not to take them at their word, since what they are asserting is only half of the modern world, the work of purification that distils what the work of hybridization supplies. (Latour 1993, 66)

Against this backdrop, Latour's own philosophical standpoint becomes increasingly clear. Latour's ontology is flat. If we grant that there is something sacred in the world, *the gods*, then there are no purely "profane" things fully detached from it, according to Latour, i.e., the divine may be associated everywhere along the network. For this reason, the *gods of today* may be found in Hölderlin's Romantic poetry, as well as in the hydroelectric plant on the banks of the Rhine, which, as we have seen in the last chapter, Heidegger pontificates as the greatest danger.

In order to better understand Latour's assessment, we need to investigate further what he means by the key concepts "purification" and "hybridization" in the quoted passage above. Purification is characteristic of the modern period and describes the process of separating knowledge from belief and superstition as well as natural things from humans. It "creates two entirely distinct ontological zones: that of human beings on the one hand; that of nonhumans on the other" (Latour 1993, 10f). Hybridization, on the other hand, designates the reverse process: it connects and mediates between separated spheres and ontological zones. This practice "creates mixtures between entirely new types of beings, hybrids of nature and culture" (Latour 1993: 10f), and is the source of the new "monsters," which we saw in the previous chapter. Without hybridization, according to Latour, "the practice of purification would be fruitless or pointless" (Latour 1993, 11). Devoid of purification, the process of hybridization "would be slowed down, limited, or even ruled out" (Latour 1993, 11). Latour maintains that the Enlightenment thinkers themselves triggered these two opposing movements:

> Freed from religious bondage, the moderns could criticize the obscurantism of the old powers by revealing the material causality that those powers dissimulate—even as they invented those very phenomena in the artificial enclosure of the laboratory. The Laws of Nature allowed the first Enlightenment thinkers to demolish the ill-founded pretensions of human prejudice. Applying this new critical tool, they no longer saw anything in the hybrids of old but illegitimate mixtures that they had to purify by separating natural mechanisms from human passions, interests or ignorance. All the ideas of yesteryear, one after the other, became inept or approximate. Or rather, simply applying the modern Constitution was enough to create, by contrast, a "yesteryear" absolutely different from today. (Latour 1993, 35)

Now that Latour's notions of purification and hybridization have been explained, we have arrived at the key ideas of Latour's philosophical project and what he takes to be the goal of his own research. Latour aims to show that the process of purification is indeed at work today and that the so-called modern movement only tried to focus on this process. By doing so, the moderns have rejected and forgotten the equally forceful and significant undercurrent of hybridization. If the interconnections between the processes of purification and hybridization, or isolation and translation , are not understood, then it is impossible to grasp the fundamental problems, paradoxes, and predicaments of what Latour calls the "modern Constitution."

> So long as we consider these two practices of translation and purification separately, we are truly modern [. . .]. As soon as we direct our attention simultaneously to the work of purification and the work of hybridization, we immediately stop being wholly modern, and our future begins to change. (Latour 1993, 11f)

In light of Latour's pejorative account of the mistakes in the modern Constitution, we need to reassess what is human and the ontology of objects. We can better position the thoughts and arguments in the first chapter as part of a more general philosophical critique of Enlightenment thought. Any clear contours between scientific knowledge and belief systems, between humans and nonhumans, disappear when we follow Latour, as they never existed in the first place: purity was merely an illusion that does not hold up to scrutiny. In place of humans and nonhumans, Latour places hybrids, quasi-objects, or *monsters*. In the first chapter, we learned how they break with the traditional categories of the moderns. Hybrids are assemblies, collectives, products of mediation, and a manifestation of the aggregated character of the world. Networks hold the world together, according to Latour—an idea which leads him to the symmetrical assessment of humans and nonhumans in what he calls a "nonmodern Constitution;" "In order to sketch in the nonmodern Constitution, it suffices to take into account what the modern Constitution left out, and to sort out the guarantees we wish to keep. We have committed ourselves to providing representation for quasi-objects" (Latour 1993, 139). And he continues, "every new call to revolution, any epistemological break, any Copernican upheaval, any claim that certain practices have become outdated forever, will be deemed dangerous, or—what is still worse in the eyes of the moderns—outdated!" (Latour 1993, 141). Latour thus seems to undermine any serious interpretation of modernity as a specific period in time, which rules out any critical account of modern technology.[3]

THE HYDROELECTRIC PLANT
AND THE GREEK TEMPLE

In Heidegger's writings, modernity imposes a great danger; it is very real and to qualify it as a mere illusion would have hazardous outcomes. Modernity, with its associated technological systems and equipment, instructs us to think about the world in functional terms: it is raw material that can be used to increase production. Modernity is the period of time in which the dangers presented in the first chapter of the book exude the greatest force. As Heidegger scholar Michael E. Zimmerman writes, "Both industrialism and modernity are symptoms of the contemporary disclosure of things as raw material to be used for expanding the scope of technological power for its own sake" (Zimmerman 1990, xiii). Inseparable from Heidegger's critical stance toward modernity and modern technology is his analysis of what he refers to as the rule of enframing. To understand his wide-ranging critique of modernity and its connection to the rule of enframing, we shall revisit his notion of technology as it is presented in "The Question Concerning Technology" (Heidegger 1977a). Following this analysis, I explain how a number of Heidegger's key insights from the domain of technology can be linked to his understanding of the work of art and his assessment of humanism. This step, moreover, prepares the discussion between Latour and Heidegger that takes place in the last part of the chapter.

The fundamental lesson to be drawn from Heidegger's study of technology is that it is not first and foremost to be understood as a specific artifact but as a way of bringing-forth beings; it actualizes new or unseen potentials in the world. By analyzing technology within the framework of bringing-forth beings, Heidegger connects technology with the ancient Greek understanding of the notion of truth, *alêtheia*. Technology as such grants access to a fundamental understanding of Being as that which precedes and renders possible a variety of specific beings. In establishing the connection between technology, *alêtheia*, and Being, Heidegger makes the assessment of technology a crucial undertaking for philosophy. In this way, Heidegger's critique of technology and modernity becomes fundamental to his own thought, just as is the case with Latour.

The ancient Greek understanding of truth reminds us that any object or any specific being must be revealed, stand out, and be intelligible before it can be analyzed in detail, for example, in the sciences, and be compared to another object (cf. also Harman 2002, 149f; Riis 2018). An object's specific identity must be revealed or brought forth before it can be determined whether the description of it is correct or incorrect. In other words, the object's identity is the result of a specific framing, which is not neutral. It should be understood as the result of a particular intervention or revealing process—a process which is also a defining criterion of the operation of technologies. These two phases—obtaining an identity and analyzing an identity—illustrate, according to Heidegger, the superiority of the ancient Greek understanding of truth over a correspondence theory of truth. He maintains that *alêtheia* describes the very revealing and rendering intelligible of beings to our experience.[4]

It is in light of this fundamental notion of technology that Heidegger distinguishes modern from ancient technology and thus launches his wide-ranging critique of modernity. Heidegger scholar Andrew Feenberg states clearly the quintessential elements of Heidegger's critique of modern technology,

> Nature is "challenged" to deliver up its wealth for arbitrary human ends. It is transformed into a source of energy to be extracted and delivered. But even as human beings take themselves for the masters of being, being "challenges" them to challenge beings [. . .] This "Gestell," this *"enframing"* within which human being and being are ordered, is now the way in which Being reveals itself. (Feenberg 2005, 21)

The main example Heidegger offers to explain the mechanisms of the "challenging revealing" of modern technology—how modern technology essentially works—is the hydroelectric plant on the banks of the Rhine. Latour also refers to this example in connection to Heraclitus as he attempts to sidestep Heidegger's notion of modernity. The Rhine revealed through modern technology becomes an energy source at human disposal. Ultimately, when nature is revealed through modern technology, it appears as what Heidegger calls "standing-reserve," or "pure stock" in the terminology of Latour. This is to say that every being, especially the human being, is considered a resource without intrinsic value; it is only significant in the endless process of generating ever more standing-reserve. Heidegger views exactly this revealing as the ultimate danger of modernity (Riis 2018).

Opposed to modern technology, Heidegger views the bringing-forth of ancient technology as a "setting free" instead of a "challenging" (Heidegger 1977a, 316). Ancient technology, in Heidegger's view, is responsible for "starting something on its way into arrival" (Heidegger 1977a, 316), and is thus less intrusive and much more sensitive to the material it is working on. In other words, ancient technology brings the proper potential or the natural telos [end] of something into appearance, whereas modern technology challenges its object to stand out and treats it like a mere resource in the endless search for more resources.

To grasp Heidegger's critique of modern technology, and find out what may save us from its alleged dangers, it is helpful to revisit his notion of the work of art. In the conclusion of "The Question Concerning Technology," Heidegger writes, "Yet the more questioningly we ponder the essence of technology, the more mysterious the essence of art becomes" (Heidegger 1977a, 341). Here, he is pointing to a fundamental affinity between the essence of technology and art and is thus alluding to his comprehensive treatment of the work of art in "The Origin of the Work of Art" (Heidegger 1977b). An in-depth analysis of Heidegger's view of the work of art is beyond the scope of this book, but I will offer a condensed version of how I interpret the connection between the two realms and explain why it is important in this comparative examination of modernity.[5]

Art is an eminent way of bringing-forth beings, which means it shares a fundamental characteristic with technology (Heidegger 1977a, 317f). The main difference between technology and art is that the event of truth inherent to bringing-forth beings is distinct as such in the artwork. Whereas the work of art is above all concerned with this phenomenon, the work of technology, which is chiefly embedded in the world of means and ends, covers up this phenomenon since it points to a comprehensive referential framework of other equipment and natural resources. In the sense that the artwork encourages insight into the very coming into being of itself, into the event of truth, and thus invites viewers to take part in understanding the precondition for any specific being, it is extraordinary. The artwork is defined by an opening that encourages its witnesses to understand the emerging of and struggle between *world* (that which is highlighted) and *earth* (that which pulls back and serves as background) (Heidegger 1977b). Technology, in opposition to the work of art, does not facilitate this insight and would not be able to function as technology if it did so. In other words, technology discourages this reflection by covering the event of truth, to which it owes its very existence.

This critique of modern technology is at the same time closely connected to the critical stance toward humanism that Heidegger developed in his *Letter on Humanism* (1946), which was written as a response to French philosopher Jean Beaufret. Heidegger was not critical toward humanism because he lacked ethical ideals; rather, he claimed humanism does not understand human beings, or *Dasein*, first and foremost in relation to the openness of Being and the event of truth, that is, in relation to their highest potential. Two main contentions from the *Letter on Humanism* show how Heidegger launches his critique: "But as long as the truth of Being is not thought, all ontology remains without its foundation" (Heidegger 1977d, 258). And this claim is closely connected to what may be seen as another programmatic statement of Heidegger: "But if man is to find his way once again into the nearness of Being he must first learn to exist in the nameless" (Heidegger 1977d, 223). Taken together, these statements present the main tenets of Heidegger's petition against humanism.[6] Humanism strives to fix or define human nature as well as a connected set of rights, duties, and ethical commands. Such a definition fails to capture the more fundamental and open understanding of being that he purposely only associates with self-reflective awareness and calls *Dasein* (being present). Connecting *human* to this being, as is the case in humanism, is a product of a *specific* understanding—something which is brought forth in a particular historical context (for that matter also *homo sapiens*). This historical process of coming into being as so-called humans is hidden by humanism in its attempt to give universal authority to various connected ethical commands.

Heidegger's interpretation and critique of humanism hence resemble his critique of technology. Humanism and technology are alike in the sense that they are taken at face value and conceal their own connection to the emergence of a specific *world* and a particular event of truth. Opposed to this, Heidegger strives to unfold the concept of the human in its historic context and openness—in its possibility structure, which allows for changes to its previous identity and destiny. In other words, *Dasein* belongs to this open realm as the *name of a nameless* being; its minimal and main feature is that it is aware of its own presence—that it is *da* (there).

In other words, the human is also a result of a *poietic* process, which we may come to understand as such by dwelling on the significance of the work of art. This connection clearly comes to the fore when Heidegger suggests a fundamental change of perspective in the relation between *Dasein* and the work of art. Such a revolution situates humans and humanism in their (open) place, as it reveals their derivative nature, which ties together the discussions above. The exemplary work of art manifesting this profound link is the Greek temple. Heidegger states:

> The temple-work, standing there, opens up a world [. . .] But men and animals, plants and things, are never present and familiar as un-changeable objects, only to represent incidentally also a fitting environment for the temple, which one fine day is added to what is already there. We shall get closer to what *is*, rather, *if we think of all this in reverse order*, assuming of course that we have, to begin with, an eye for how differently everything then faces us [. . .] The temple, in its standing there, first gives to things their look and to men their outlook on themselves. (Heidegger 1977b, 168; emphasis added)[7]

We have now come closer to understanding the profound connection between Heidegger's assessment of technology, artworks, and his considerations concerning humanism. Based on this connection, this Ariadne thread, we can examine the important commonalities and conflicts between Latour and Heidegger, which will serve as the basis for the discussion in the third and final part of this chapter.

NETWORKS, ARTWORKS, AND TECHNOLOGIES

Stated in its most radical form, Latour accuses Heidegger of being a modernist. Latour charges him with believing in the modern Constitution—although as a threat, but nonetheless. It is in the context of this unfavorable assessment of Heidegger that Latour also calls attention to Lévi-Strauss and suggests that Heidegger even provokes and enforces the development that he warns us about. In Latour's words:

> As Lévi-Strauss says, "the barbarian is first and foremost the man who believes in barbarism." Those who have failed to undertake empirical studies of sciences, technologies, law, politics, economics, religion, or fiction have lost the traces of Being that are distributed everywhere among beings. If, scorning empiricism, you opt out of the exact sciences, then the human sciences, then traditional philosophy, then the sciences of language, and you hunker down in your forest—then you will indeed feel a tragic loss. (Latour 1993, 66)

Keeping in mind this accusation aimed at Heidegger and the above interpretations, we shall now proceed to answer the question asked in the opening passage of this chapter: Have we never been modern? My claim is that Latour and Heidegger principally agree about the potential threat of the modern Constitution, but that Latour fails to understand[8] the crucial link between the threat of modern technology and the work of art, which Heidegger mentions (but does not unfold in detail) in "The Question Concerning Technology."[9] From this perspective, the modern Constitution may indeed be seen as what I have elsewhere described as a temporal artwork: a way of shaping and framing the unfolding of time and turning it into a specific epoch of Being (Riis 2018). The rest of the chapter will discuss the fundamental connection between Heidegger's notion of an artwork and Latour's understanding of networks. Both will be discussed with regard to their implications for the constellation of humans and technologies and the notion of modernity.

To support the claim concerning the fundamental commonalities of Heidegger's and Latour's assessments of modernity and the modern Constitution, I will argued that Latour, in a specific yet *backward* way, explicitly affirms Heidegger's critique of modernity and the associated dangers. In a significant paragraph of *We Have Never Been Modern*, Latour writes, "Who has forgotten Being? No one, no one has, otherwise Nature would be truly available as pure 'stock'" (Latour 1993, 66). Here, Latour, in fact agrees with a crucial part of Heidegger's concern, that is, if Being as the potentiality of beings has been forgotten, then Nature and all beings actually will appear as pure stock, or, what is the same, as standing-reserve. The resulting question then becomes how it is possible to find out whether Being in the Heideggerian sense has truly been forgotten. And it is this crucial question that Heidegger and Latour seem to disagree about in their responses.

As we have seen above, in order to understand Being in Heidegger's terms, we need to understand the process by which something specific appears to us (*alêtheia*), and thus how revealing and concealing are inseparable parts of the process which reveals beings and makes them comprehensible to us: they determine how a specific object *faces* us, for example, which parts of it, its surroundings, or parts of our own thinking process are being excluded, overshadowed, hidden, or neglected. Being in this sense is never exhausted by any particular beings but marks the potentiality of what may come into existence in the future. This is what Heidegger paradigmatically calls attention to in the work of art when he writes about the struggle between earth and world (cf. Heidegger 1977b).

Based on Heidegger's insight and translated into the specific context of the apparent tension between Heidegger's and Latour's respective positions on modernity, it is possible to defend the initial claim of this chapter by calling attention to *alêtheia* as the sovereign which determines the threads and actors of the network that stand out, which knots are tied, and which threads and actors are overlooked. In Latour's notion of the network, *alêtheia* manifests specific ties and associations between beings in the world, as well as their interdependence and disposition. The artwork, in Heidegger's terms, manifests the same rule over appearance in the sense that it decides what is perceived and what recedes into the background. Or better yet, works of art make it possible for us to understand the logic governing revealing and concealing, even in Latour's networks. In other words, Heidegger's notion of art is not only compatible with Latour's idea of a network between beings but suggests such a network.

If the structure of revealing and concealing—the struggle between earth and world—is immanent to the appearance of every being, of every*thing* and every process of bringing-forth, then we may also better understand how new ideas, concepts, and ideals pontificate certain aspects, actors, ties, and relations in the world, while downplaying, isolating, ignoring, and neglecting others. Just as a painting brings certain beings into perspective and draws attention to a particular stance in the visual realm, so words and concepts highlight certain ideas in the intellectual realm. Based on the latter assessment, we may indeed also learn to see how a specific revealing process belongs to the modern Constitution and the Enlightenment project. The modern Constitution is a constitution in the sense that it enshrines a certain account and view of the world. Heidegger acknowledges this when he associates the work of truth not only with the work of art but also with the constitution of a state, "One essential way in which truth establishes itself in the beings it has opened up is truth setting itself into work [the artwork].

Another way in which truth occurs is the act that founds a political state" (Heidegger 1977b, 186).[10]

The praxis of purification, which Latour closely associates with the modern Constitution, is to be understood as a real intellectual and empirical revealing process: it discloses a specific view of the world, of the sciences, and of ourselves in a process that links powerful networks. A constitution is influential in the sense that it might determine the identity of objects; what is associated and tied together; what stands out and what is ignored; and what is to be regarded as attractive and repulsive and as good and evil. It also "gives to things their look and to men their outlook on themselves" (Heidegger 1977b, 168), as Heidegger writes in "The Origin of the Work of Art." In this way, we can learn to *see* how exactly a constitution has *constitutive* power, how it creates tensions, and how it generates allies and enemies, which again produce certain struggles and particular kinds of connections and ties.

Based on this interpretation, we shall now return to Latour and my claim that the modern Constitution is no mere illusion. The modern Constitution carries out its thorough work by means of—among others—an immense workforce of scientists, engineers, and politicians; its vast variety of images and representations of the world; and its educational system, which shapes the future workforce (Hilt 2005; Balslev 2018; Riis 2018). In this sense, there are many moderns, and their number might even be growing. Latour would most likely agree with this statement, as he himself ventured into the so-called Science Wars of the 1990s and eagerly fought against what he took to be the allies of the modern Constitution. As real as this war was, just as real were the opponents. In an imaginary dialogue entitled "Science Wars," Latour plays out the strong accusations between a representative of the *modern* concept of science (*she* in the dialogue below), and the position Latour defends (represented by *he* in the dialogue below). As this dialogue unfolds, Latour writes:

> *HE:* Very well, if you wish: I'm a relativist in the sense that I, like you, reject an absolute point of reference. I agree that this rejection permits me to establish relations and distinctions, and to measure the gaps between points of view. For me, being a relativist means being able to establish relations between frames of reference, and so, being able to pass from one framework to another in converting measurements (or, at least, explanations and descriptions). It's a positive term, I agree, to the extent that the opposite of *relativist* is *absolutist*.
>
> *SHE:* If what you say is true, why do my colleagues so attack you? Are you keeping something from me? You are a wolf in sheep's clothing, *n'est-ce pas?*
>
> *HE:* Forgive me, but your colleagues aren't simply physicists, they're politicians too, and it's for *political* reasons that they call me every name under the

sun. They're wolves pretending to be sheep under attack by wolves. (Latour 2002b, 72)

In other words, in order to fight this war, you need to believe in and take seriously the effect of the modern Constitution and what the moderns think about themselves. You might need to become a "barbarian" in the sense of Lévi-Strauss to see this—or maybe a hybrid Latourian monster, such as a sheep in wolf's clothing.

Latour highlights the importance of hybridization and mediation as expressions of the network, as well as the reciprocal influence between things, but this does not mean that he does not believe the *danger* of the modern Constitution is real; on the contrary. Hybridization and mediation are the conceptual weapons or tools that he utilizes to generate a different picture of the sciences in the Science Wars. The phenomena of mediation and hybridization are indeed also closely connected to Heidegger's interpretation of the work of art, his notion of revealing, and the modern Constitution: without mediation, there would be neither a way to exercise the rule Heidegger ascribes to the work of art, which links beings in new ways, nor a pathway for the purification process to take place. And without co-constitution among the elements of an artwork or a constitution, one could not even speak of an artwork or a nonmodern constitution for that matter.

Even though purification and mediation are opposing processes, they are in another way preconditions for each other. Each is the flipside, or the *Kehrseite*, of the other, as Heidegger would say. In the Science Wars, Latour emphasized that there is mediation taking place everywhere there is purification (Latour 1993, 35). That is to say, an undercurrent runs below the purification processes—the former is ignored by the modern Constitution and the latter is highlighted. It is this dialectical insight and assessment of Latour that we may now understand better:

> So long as we consider these two practices of translation and purification separately, we are truly modern—that is, we willingly subscribe to the critical project, even though that project is developed only through the proliferation of hybrids down below. As soon as we direct our attention simultaneously to the work of purification and the work of hybridization, we immediately stop being wholly modern, and our future begins to change. At the same time, we stop having been modern, because we become retrospectively aware that the two sets of practices have always already been at work in the historical period that is ending. Our past begins to change. (Latour 1993, 11)

In other words, Latour claims that there is more to the modern Constitution than purification; that is, purification is always accompanied by mediation, and this insight is going to save us from the promises and perils of purity. On the face of it, the modern Constitution preaches purification, but purification relies on hybrid laboratories and new technologies and conceptual constructs.

Based on Latour's arguments, it could be argued that we have never been exclusively modern if we understand this to mean that there has never only been purification. However, Heidegger would agree with Latour in this respect, and he argues exactly in the same vein with respect to modernity. Heidegger also points out that at the moment of despair, when the production of standing-reserve overwhelms us, we need to understand the flipside and acknowledge the fundamental revealing processes taking place where the standing-reserve is ruling. Production processes do not belong to enframing alone but are dependent upon the more primordial event of truth, which is most clearly manifest in the work of art. In other words, modern technology is not a *pure* offspring of enframing but a close relative to art and the event of truth. Modern technology cannot be overcome by ignoring technology or trying to stop its advance. On the contrary, we can only be saved from its dangerous influence and not turned into standing-reserve by paying close attention to how it really operates and creates a certain constellation of concealing and disclosing Being. This is the epiphany of the artwork. We may thus learn to see that technologies proliferate in a certain epoch, which is termed modernity. In another more fundamental sense, modern technology and the rule of enframing belong to and are dependent upon the nonmodern revealing process, *alêtheia*, which is ubiquitous. In this sense, the truth sets free the rule of enframing, according to Heidegger.

Latour, like Heidegger, recommends not taking the modern Constitution or modern technologies at face value but investigating these constructs in order to discover what is really at stake in their emergence and under their rule (Riis 2018). Just as *alêtheia* and the struggle between revealing and concealing and the war between purification and mediation do not belong specifically to modernity, we also need to see that humanism and its associated constructs of humans and nonhumans are historically emergent phenomena that owe their existence to the process of *alêtheia* and the openness of Being. So, according to both Heidegger and Latour, we have never been exclusively modern, but the modern Constitution and the rule of enframing fundamentally influence beings in the world today.

Finally, we may now also address the *we* in the initial question above: "Have We Never Been Modern?" Who is "we" in this question? "We" stands for, first and foremost, the beings exposed to Being who are dependent upon *alêtheia*, and thus also subject to change, new epochs, and the dangers of the modern Constitution—a danger that Heidegger counters by making us aware of a more fundamental constitution, namely, the constitution of Being. Latour tries to save us from the same danger by revealing a prior, timeless, and nonmodern Constitution of Being. However, as we have seen above, both thinkers take seriously the modern revealing of the world as a danger we cannot afford to ignore—a danger that we, as beings associated with truth, are not at the mercy of but may subversively bypass. How this might work will be studied in further detail after another ontological and existential link between Latour and Heidegger has been investigated. In the next chapter, the alleged symmetry between humans and nonhumans undergoes an untraditional existential analysis that emphasizes what is at stake in the controversy between Latour and Heidegger.

NOTES

1. Cf. chapter 6.
2. In Latour's latest writings, he seems to change that view quite radically, cf. chapter 6.
3. In Latour's latest writings, his views do change in light of the unfolding environmental disasters. See also chapter 6.
4. It is this process that phenomenology seeks to trace (cf. chapter 5).
5. Cf. Riis 2018.
6. Cf. Till (2004).
7. Jacob Wamberg's comprehensive and highly interesting work can be seen to support this argument made by Heidegger. Wamberg has studied landscape art since its origin and shown how our view of landscapes is co-shaped by the kind and degree of technological advancement (Wamberg 2009).
8. Alternatively, based on David Bloor's critique of Latour, it is also possible to see a systematic misrepresentation of Heidegger in Latour's writings: "I shall conclude that [Latour's] criticisms are based on a systematic misrepresentation of the position he rejects [. . .]" (Bloor 1999, 82).
9. Cf. Riis (2018).
10. Cf. Jasanoff and Kim (2015).

Chapter 3

Death and the End of Networks

Heidegger unfolds his concept of Dasein in *Being and Time*, which was brought to our attention in the previous chapter. This notion describes the foundational structure and critique of the human mode of being. For Heidegger, humans are always related to the world; that is, human existence is always to be understood in relation to the world. Heidegger first shows this decisive relation between Dasein and the world in view of the revealing of the world, which was exemplified in his interpretation of the artwork but is also manifested in the everyday practical character of human life and in human understanding. He expresses the inner relation between Dasein and the world with his well-known formula of the foundational constitution of Dasein: being-in-the-world.[1]

Latour's critique of the so-called modern Constitution articulates the far-reaching implications of the simultaneous birth of humans and '"nonhumanity'—things, or objects, or beasts" (Latour 1993, 13). Based on this basic *symmetry* between humans and nonhumans, Latour develops his Actor-Network-Theory (ANT), which was crafted to analyze the different ways in which humans and nonhumans are intertwined in everyday life:

> If anything, the modern collective is the one in which the relations of humans and nonhumans are so intimate, the transactions so many, the mediations so convoluted, that there is no plausible sense in which artifact, corporate body, and subject can be distinguished. (Latour 1999, 197)

Heidegger and Latour share several research interests and have developed a number of related concepts. However, it is just as important to understand how they differ in terms of the scope and intimacy of the fundamental relation between humans and the world. When taking these aspects into consideration, their positions are in absolute disagreement.

According to Heidegger, Dasein is embedded in the surrounding world, and it is this existential entanglement of Dasein with the world that makes it difficult to draw an exact line between Dasein and its environment, and thus it is also not easy to understand Dasein independently of the things in the surrounding world. At the same time, it is important to note that a constitutive trait of Dasein is its consciousness of its own finitude. Reflecting on the temporal boundary of death becomes a decisive marker of the fundamental difference between Dasein and other beings; even more, death in Heidegger's thought is an indication of the principal limitation and peculiarity of Dasein.

In contrast, Latour sees no ultimate challenge to the symmetry between humans and nonhumans, and he pleads for the infiniteness of networks, which connect humans and nonhumans and eliminate their individuality. In Latour's conception, every *negative conception* is relative; all limits and boundaries can be countered with boldness and shrewdness, and death is either ignored or seen through the perspective of the shackling networks. Latour focuses on movement, metamorphosis, and life—he allows no room for stillstand, silence, or death.

This difference between Heidegger and Latour enables a new Heidegger-inspired critique of Latour's ANT and his late *magnum opus* titled *An Inquiry into Modes of Existence* (*AIME*). By the same token, it is possible to uncover fundamental pantheistic conceptions in Latour's networked thought.[2] In other words, by considering the Heideggerian existential analytic, the aim of this chapter is to develop a contemporary critical response to ANT. This Heidegger-inspired critique, on the one hand, unearths an overlooked risk of ANT and, on the other, suggests the idea of death and finitude as constitutive moments of a social ontology of Dasein.

BEING-IN-THE-WORLD AND ITS LIMITATIONS

To understand the relation between Dasein and the world in Heidegger, it is important to consider the basic structure of Dasein: being-in-the-world. From this formula, it is apparent that Dasein is fundamentally embedded in the world, but not in the sense of an isolated atom surrounded by a world. And this is exactly one of the most basic misunderstandings which Heidegger seeks to prevent by using hyphens. With this formula, Heidegger points to how Dasein is always already revealed in a world and that it is related to it in its mode of being.

> What is meant by *"Being-in"*? Our proximal reaction is to round out this expression to "Being-in 'in the world,'" and we are inclined to understand this Being-in as "Being in something" [*"Sein in . . ."*]. This latter term designates the kind of Being which an entity has when it is "in" another one, as the water is "in" the glass, or the garment is "in" the cupboard. By this "in" we mean the relationship of Being which two entities extended "in" space have to each other with regard to their location in that space [. . .] Being-in, on the other hand, is a state of *Dasein*'s Being; it is an existentiale. So one cannot think of it as the Being-present-at-hand of some corporeal Thing (such as a human body) "in" an entity which is present-at-hand. Nor does the term "Being-in" mean a spatial "in-one-another-ness" of things present-at-hand, any more than the word "in" primordially signifies a spatial relationship of this kind. "In" is derived from "innan"—"to reside," "habitare," "to dwell" [*sich aufhalten*]. "An" signifies "I am accustomed," "I am familiar with," "I look after something" [. . .] "Being," as the infinitive of "*ich bin*" (that is to say, when it is understood as an *existentiale*), signifies "to reside alongside . . .," "to be familiar with . . ." "*Being-in*" *is thus the formal existential expression for the Being of Dasein, which has Being-in-the-world as its essential state.* (Heidegger 2008, 54)

Dasein is originally related to the world, and the intentional character of its Being is manifested in the familiarity of everyday things (Khong 2003, 698). According to Heidegger, Dasein is thrown into the world, that is, it is there (*da*) and *is* in a fundamental sense related to the world.[3] This is why beings in Heidegger's ontology are initially conceived as part of a concrete, lifeworld *praxis*. Beings are not to be considered, initially, from a distance as abstract entities and imagined as if Dasein has no contact with them—quite the contrary; between Dasein and beings exists a basic compatibility which Heidegger expresses as "readiness-to-hand" (*Zuhandenheit*). Things are, in other words, understood first and foremost in relation to "hand," that is, in relation to praxis, before they can be conceived as individual objects outside of everyday use.[4] The relations between things are uncovered and understood through their "involvement" (*Bewandtnis*). In the resulting praxis-based referential framework of beings, all beings obtain their meaning by means of their function in the encompassing use-framework. The concept of beings as abstract, existing outside of the referential framework—as merely "present-at-hand"—was developed subsequently.

> The kind of being which belongs to these embedded entities is readiness-to-hand. But this characteristic is not to be understood as merely a way of taking them, as if we were talking such *aspects* into the *entities* which we proximally encounter, or as if some world-stuff which is proximally present-at-hand in itself were *given subjective colouring* in this way. Such an interpretation would overlook the fact that in this case these entities would have to be understood and discovered beforehand as something purely present-at-hand, and must have

priority and take the lead in the sequence of those dealings with the *world* in which something is discovered and made one's own. However, this already runs counter to the ontological meaning of cognition, which we have presented as an essential mode of Being-in-the-world. To lay bare what is just present-at-hand and no more, cognition must first be familiar with and look beyond what is ready-to-hand in our concern. "*Readiness-to-hand is the way in which entities as they are 'in themselves' are defined ontologico-categorially.*" (Heidegger 2008, 71)

To understand Dasein's constitution, it is just as important to understand that it does not entirely disappear in its daily routines and practicalities, since Dasein is a temporal being in the sense that it is conscious of its past and present and stands in relation to its future possibilities. Due to Dasein's temporal structure, it is never completed, but also to be understood in the context of its not-yet, that is, its death.[5]

Death is a possibility-of-Being which *Dasein* itself has to take over in every case. With death, *Dasein* stands before itself in its ownmost potentiality-for-Being. This is a possibility in which the issue is nothing less than *Dasein*'s Being-in-the-world. Its death is the possibility of no-longer being-able-to-be-there. If *Dasein* stands before itself as this possibility, it has been *fully* assigned to its ownmost potentiality-for-Being. When it stands before itself in this way, all its relations to any other *Dasein* have been undone. This ownmost non-relational possibility is at the same time the uttermost one. As potentiality-for-Being, *Dasein* cannot outstrip the possibility of death. Death is the possibility of the absolute impossibility of *Dasein*. Thus death reveals itself as *that possibility which is one's ownmost, which is non-relational, and which is not to be outstripped* [*unuberholbare*]. (Heidegger 2008, 251)

Because Dasein is conscious of its own end, it is in a unique position to understand itself as utterly "non-relational." The finitude of Dasein is not only to be viewed negatively, but it manifests itself through different kinds of examination of the world; the finitude of Dasein can also be experienced through the radical form of anxiety, or *Angst*. But, first, the phenomenon of death must be understood better, as it will play a key role later in the critical assessment of ANT.

REPRESENTABILITY, FINITUDE, AND DEATH

As I have shown above, Dasein is fundamentally characterized through its being-in-the-world and is thus directed at the world; it is initially understood in accordance with its functions in everyday practical life.[6] In Heidegger's interpretation of Dasein, it is crucial to understand that its practical activity can be represented by others, "In everyday concern, constant and manifold use is made of such representability. Whenever we go anywhere or have anything to contribute, we can be represented by someone within the range of that 'environment' with which we are most closely concerned" (Heidegger 2008, 239).[7] Because Dasein understands itself partly in terms of the tasks it carries out, there emerges a kind of *über*-individual identity, which Heidegger calls the *das Man*, which is translated interchangeably into English with the impersonal "one" or "they": "But *proximally and for the most part* everyday *Dasein* understands itself in terms of that with which it is customarily concerned. 'One is' what one does" (Heidegger 2008, 239; emphasis added).

"Representability" (*Vertretbarkeit*) is an essential characteristic of Dasein and concerns both the general, open mode of being of Dasein and the specific mode of being: occupation, social status, or age.[8]

> We take pleasure and enjoy ourselves as they [*das Man*] take pleasure; we read, see, and judge about literature and art as they see and judge; likewise we shrink back from the "great mass" as they shrink back; we find "shocking" what they find shocking. The "they," which is nothing definite, and which all are, though not as the sum, prescribes the kind of Being of everydayness [. . .] Everyone is the other, and no one is himself. (Heidegger 2008, 126)

The collective and anonymous identity of *one* is also representable in *one's* profession, and this is expressed in particular with the idea of the "functionary."[9] The functionary performs a function that another person can carry out or learn to do just as well. After the industrial revolution and above all in the present age through different technologies, the functionary has become replaceable. Because *Dasein proximally and for the most part* is understood in terms of its everyday activities—which can be performed by another, or at the beginning of the twenty-first century by sophisticated new technology—representability provokes a very instrumental interpretation of Dasein. According to this image, the everyday activity of Dasein is the focus insofar as it is representable by others or new technologies. Representability—by *the they* and reinforced by technology—can be seen as evidence of a *symmetric ontology*, which Latour also highlighted, and it is characterized by the functional egalitarianism of humans and nonhumans, that is, technologies and things in general.

Representability ends, however, for Heidegger, when one is confronted with death. It can appear as if death is also representable by other humans. Dasein can experience death by attending the death of another, how he or she stops being-in-the-world, or stops being *there*.[10] It can sacrifice itself for another, but in doing so it does not do away with the inevitability of death:

> However, this possibility of representing breaks down completely if the issue is one of representing that possibility-of-Being which makes up *Dasein*'s coming

to an end, and which, as such, gives to it its wholeness. No one can take the Other's dying away from him [...] In "ending," and in *Dasein*'s Being-a-whole, for which such ending is constitutive, there is, by its very essence, no representing. (Heidegger 2008, 240)[11]

Based on the principle and absolute unrepresentability of death, the life of Dasein changes: it is individualized in terms of its end.[12] Ultimately, for Dasein, death means that its life is fundamentally defined by "mineness" (*Jemeinigkeit*).[13] Stated in other terms, because death is absolute for Heidegger, it can individualize life. Death leads to Heidegger's paradoxical summary of the existence of Dasein as the "coming-to-its-end of what-is-not-yet-at-an-end" (Heidegger 2008, 242).[14] In this fundamental and permanent relation to death (and anticipation of death), the existence of Dasein unfolds and in doing so demarcates itself from other beings. "The 'ending' which we have in view when we speak of death, does not signify *Dasein*'s Being-at-an-end [*Zu-Endesein*]" (Heidegger 2008, 245), according to Heidegger, "but a Being-towards-the-end [*Sein zum Ende*] of this entity. Death is a way to be, which *Dasein* takes over as soon as it is" (2008, 245). In other words, *Dasein qua Dasein* knows that it is sentenced to death, so to speak, and this fact influences its life radically. With its imminent, absolute, and unsurpassable end and limitations, Dasein searches either for the mindless life of "the they" or actual existence (Heidegger 2008, 250).[15] But in both modes, death is inevitable and continually reveals the limitations of Dasein—and as such, death is decisive for both modes of being. As the limits of its own existence and the negation of Dasein, death is ultimately to be understood as a negative principle—and the negativity of death is the condition of every relation to other beings.

FALLING AND ASCENT

For Heidegger, Dasein flees first and foremost when countering its own non-referencing, inevitable possibility of existence, which is revealed to Dasein in the phenomenon of death. But exactly this fleeing—the covering of uncovered possibilities—is evidence of the decisive meaning of death for Dasein in the mode of being of the they and even suggests fleeing indirectly. Heidegger refers to this fleeing as a *falling*; it can also be seen as recourse to the average life of the they.[16]

> But in thus falling and fleeing *in the face of* death, *Dasein*'s everydayness attests that the very "they" itself already has the definite character of *Being-towards-death*, even when it is not explicitly engaged in "thinking about death." Even in average everydayness, this ownmost potentiality-for-Being,

which is non-relational and not to be outstripped, is constantly an issue for *Dasein*. This is the case when its concern is merely in the mode of an untroubled indifference towards the uttermost possibility of existence. (Heidegger 2008, 254)

Falling manifests itself in a series of different phenomena, which Heidegger analyzes abstractly, yet quite forcefully.[17] To prepare the discussion of the similarities and differences of Heidegger and Latour with respect to being human, we must first consider falling in the form of curiosity.

One mode of falling is the characteristic non-lingering, which is distinctive of curiosity. In the continual distraction pertaining to curiosity, Being escapes the gravity of mineness, and in this way, novelty and the anticipation thereof actually block the way to Being. Curiosity "seeks the leisure of tarrying observantly, but rather seeks restlessness and the excitement of continual novelty and changing encounters" (Heidegger 2008, 172). This non-lingering of curiosity is apparent in the many present-day organizations and institutions focused on innovation. Based on Heidegger's Dasein analytic, the current drive for innovation can be understood as a continual incitement of curiosity—a way in which curiosity keeps itself in gear and perpetuates the falling.[18] Research aiming to bring about new innovations or analyzing existing innovations to the last detail would be, for Heidegger, incapable of advancing to the primordial connection with the existence of Dasein.

For Heidegger, Dasein, however, can be abruptly torn from curiosity and the mode of being of the they. This happens in the phenomenon of anxiety. The experience of anxiety momentarily ends the falling. Heidegger writes, "In anxiety what is environmentally ready-to-hand sinks away, and so, in general, do entities within-the-world. The 'world' can offer nothing more, and neither can the *Dasein*-with of Others" (Heidegger 2008, 187). In the experience of anxiety, there is no occasion for curiosity, that is, the excitement connected to novelty, and that which is ready-to-hand disappears. In anxiety, Dasein is confronted with the question about the meaning of the whole when it turns toward novelty:

Anxiety thus takes away from *Dasein* the possibility of understanding itself, as it falls, in terms of the "world" and the way things have been publicly interpreted. Anxiety throws *Dasein* back upon that which it is anxious about—its authentic potentiality-for-Being-in-the-world. Anxiety individualizes *Dasein* for its ownmost Being-in-the-world, which as something that understands, projects itself essentially upon possibilities. Therefore, with that which it is anxious about, anxiety discloses *Dasein* as Being-possible, and indeed as the only kind of thing which it can be of its own accord as something individualized in individualization [*vereinzeltes in der Vereinzelung*]. (Heidegger 2008, 187)

This "existential solipsism"[19] should, however, not be misunderstood: Dasein is not an isolated subject. In the experience of anxiety, the being-in-the-world of Dasein is confirmed; the world is not merely these or those beings, but it is the world as such into which Dasein was thrown and must rely on.[20] The negative and concurrent positive experience of the world in anxiety establishes for Heidegger the exceptional mode of being of Dasein. The distinguishing characteristic of Dasein is that it has a world and that it is the only entity that can fall through the world but also come to know the disclosedness of the world as such. Heidegger follows these thoughts about disclosedness further by connecting them with the concept of truth. Truth here is understood in accordance with the ancient Greek notion of uncoveredness, *alêtheia*, as emphasized in previous chapters: "[A]ll uncoveredness is grounded ontologically in the most primordial truth, the disclosedness of *Dasein*. As an entity which is both disclosed and disclosing, and one which uncovers, *Dasein* is essentially 'in the truth'" (Heidegger 2008, 256).

Even if Dasein first and foremost shows itself in the mode of the they, and it appears in average everydayness, it is to be understood in terms of its mode of being-in-the-world. Dasein can understand this mode of being as such, as well as the extent to which it can exist *in truth*. The ordered life of the they in mass society may have similarities with ants following an ant colony on the beaten track, but Dasein is, in Heidegger's conception, radically different from the mode of being of other beings. Dasein is *in the truth* and can relate to its "ownmost potentiality-for-Being, which is non-relational and not to be outstripped" (Heidegger 2008, 254). The life of the they is, against this background, only understandable as a fleeing, and can merely be understood from now on as a kind of falling. Even if the they appears to be the primary mode of Dasein, it is only a secondary mode of Being in Heidegger's Dasein analytic. Against this background, we can now turn to Latour and see how he explicitly enthrones Heidegger's notion of the "inauthentic existence" of human beings.

THE INFINITE NETWORK OF LATOUR

Similar to Heidegger, Latour emphasizes the close relation between humans and their environment, that is, between humans and things, technologies, animals, and objects of all sorts. In Latour's view, the world has been colonized by innumerable humans and nonhumans, which, due to day-to-day dealings, cooperate with each other in their various business affairs. Humans could simply not exist and function without their better halves, the nonhumans. To emphasize *a priori* the commonalities of humans and nonhumans from the point of view of ANT, Latour uses the term "actor" or "actant" to refer to both in a nondiscriminatory way,

> Action is simply not a property of humans but of an association of actants [...] Provisional *actorial* roles may be attributed to actants only because actants are in the process of exchanging competences, offering one another new possibilities, new goals, new functions. (Latour 1999, 182)

Latour is interested in humans and in nonhumans to the extent that they are active. Latour depicts the activities of humans and nonhumans as connected, transferred, controlled, and rearranged by an overarching network.[21] This view of the entanglement of humans and nonhumans is also illustrated on the book jacket of Latour's first edition of *Reassembling the Social: An Introduction to Actor-Network-Theory* (2005). Shown here is the lithograph by Bompied entitled "Exposition de Madagascar: le panorama de la 'Prise de Tananarive' pendant son execution" (1900). It portrays humans cooperating with nonhumans (cranes, steel constructions, and a train with movable parts) to *conquer* Madagascar. Hinted at here is the actual conquest of Madagascar by a leading industrial nation, as well as the L'Exposition Universelle in Paris (1900). The artist painting the entire panorama also appears in the lithograph. Through this classical motif, the entire picture produces a self-reflective effect. In other words, we are looking at a representation of the representers of a representation, which puts the entire image in a new light and, at the same time, informs us visually of Latour's convoluted relational sociology.

We can better understand Latour's idea of actors and networks if we recognize that ANT is not merely to be understood as an abbreviation but also literally as the uninterrupted work of ants in higher services: "The acronym A.N.T. was perfectly fit for a blind, myopic, workaholic, trail-sniffing, and collective traveler. An ant writing for other ants, this fits my project well" (Latour 2005a, 9). This self-reflexive interpretation of Latour's work can be seen as a summary of what is expressed in several of his books, but which is especially prevalent in *Aramis or the Love of Technology*.[22] In this book Latour unfolds in detail the immense *net-work* of various actors (actants) who worked together to plan and attempt the execution of a visionary personal transit project in Paris called Aramis. The network of this project included thousands of different actors who worked, interacted, and communicated with each other—each one like a small ant participating in the construction of a gigantic ant hill. Humans and nonhumans were indispensable for the work on this project, and each actor had their own function to fulfill and instructions to follow. In his book, Latour animates the technology in question by simulating the voice of the never-entirely-completed Aramis project.

> The motor breaks down, the onboard steering shakes and shatters, the automatic features are still heteromats overpopulated with people in blue and white smocks. Chase away the people and I return to an inert state. Bring the people back and I am aroused again, but my life belongs to the engineers who are pushing me, pulling me, repairing me, deciding about me, cursing me, steering me. No, Aramis is not yet among the powers that be. The prototype circulates in bits and pieces between the hands of humans; humans do not circulate between my sides. I am a great human anthill, a huge body in the process of composition and decomposition, depending. (Latour 1996, 123)

In agreement with the ANT understanding of humans and nonhumans, Latour develops a concept of corporate bodies, a collective that decides about the coordination and direction of the actors' acting.

> Purposeful action and intentionality may not be properties of objects, but they are not properties of humans either. They are the properties of institutions, of apparatuses [. . .] Only corporate bodies are able to absorb the proliferation of mediators, to regulate their expression, to redistribute skills, to force boxes to blacken and close. Objects that exist simply as objects, detached from a collective life, are unknown, buried in the ground. (Latour 1999, 193)[23]

In Latour's ontology, all beings possess a kind of agency and stand in relation to each other. Although the connections can sometimes be very complex and intertwined, there is always a thread in the network connecting them. This we may understand similarly to the theory of six degrees of separation, in which all humans are connected to each other in a network through six persons.[24] According to Latour, all humans and nonhumans are connected and oftentimes through far fewer links than six. For instance, many humans come into contact with each other through a computer; they depend on this machine, whose parts stem from raw materials all over the world and form a global logistics network.

> Yet there is an Ariadne's thread that would allow us to pass with continuity from the local to the global, from the human to the nonhuman. It is the thread of networks of practices and instruments, of documents and translations. An organization, a market, an institution, are not supralunar objects made of a different matter from our poor local sublunary relations [. . .] The only difference stems from the fact that they are made up of hybrids and have to mobilize a great number of objects for their description. (Latour 1993, 121)

In Latour's world, no actor is lonely or could be experiencing *anxiety*; the actors are always articulated in comprehensive and sometimes infinite networks connected to each other, ordered and plugged in. Individuality is a modern illusion. Humans and nonhumans have the same potential to contribute to the network, that is, to realize the intention of the collective body. It is, however, not possible for them to resist the intention of the collective altogether.

THE TWOFOLD EVASION OF DEATH

The omnipresent network connecting all humans and nonhumans, shedding new light on their individuation, has radical consequences for the understanding of death. Although Latour does not focus on these implications explicitly in ANT—his texts show no explicit examination of death—we can look to his categorical network-thoughts and different passages in his texts for an implicit understanding of death.

When humans die, or nonhumans are destroyed or stop working as they did before, their agency, for Latour, *does not stop*. Considered as objects, they merely become parts of new networks with different functions: the (no-longer) human who has been buried could now play a new role, for instance, in a complicated underground environment of bacteria, wood, and dirt. And it is likewise the case with nonhumans, such as garbage that has been thrown away—they could also become a part of such a underground network. If one follows these actors, which Latour calls upon his readers to do, then one recognizes that in death there cannot be a radical caesura, much less an end of agency. According to ANT, we should see rather continuity and transformation in action instead of discontinuity—there can be a shift in the drive of actors, but if we look at the various processes, there are no absolute breaks but associated transformation, translation, hybridity, and construed continuities.[25] The network's drive remains intact just as energy does in Newton's mechanistic cosmology. And this is exactly the point: movement in Latour's thought is at the center and has no end, but it may manifest itself differently and change direction.

Based on these considerations of the omnipresent energy in the world of Latour, steered by global networks, it is possible to recognize in his thought a new articulation of pantheism.[26] In Latour, there are more or less autonomous collectives providing all the things they network with movement and energy, which give them a more or less connected orientation.[27] The collective outlasts the individual things and individual humans that it connects. In other words, the death of each individual is a brief, cursory, unimportant phenomenon, and it posits no absolute caesura in the network—the network is, as before, animated from energy.[28] In the preliminary study for his ambitious book *AIME*, Latour wrote "Reflections on Etienne Souriau's *Les differents modes d'existence*" (2011). Latour describes in this work the particular mode of existence of souls. He claims, "'Having a soul' is a task to be accomplished, and it can be botched and most often is" (Latour 2011, 321). Latour elaborates that he is not talking about an immortal substance—it can be lost but also brought back once again (Latour 2011, 321). It is important to note here that Latour does not connect the loss of the soul with death in the sense of Heidegger. There are no implications between the two.

A detailed analogy of Latour's articulations in ANT, *AIME*, and pantheism is beyond the scope of this chapter. However, to make this analogy more understandable, it should be emphasized that the foundational idea of pantheism is expressed briefly and precisely in Spinoza's principle: "God is the immanent but not transitive cause of all things" (Ethics 1, P18). The movement, the action, the energy which is omnipresent and immanent in Latour's network can be understood in agreement with Spinoza's doctrine, and at the same time as a kind of "world-soul," which is a foundational concept by another pantheistic philosopher, namely Giordano Bruno. He writes about the world-soul,

> Therefore, the Soul of the World is the formal constitutive principle of the universe, and of whatever the universe includes; I mean, if life is found in all things, then the soul is the form of all things; it controls matter in all respects and is prevalent in the compounds, operates the composition and the consistency of the parts. (Latour 1977, 39)

In Bruno's thinking, there is no division between this world and the afterworld—the divine principle is in all things, but still thought of as its own principle.

In what Latour presented as his "Spinoza Lectures" at the University of Amsterdam he builts on Alfred Whitehead and delivers a polemic against the bifurcation of nature and advances the pantheistic thought a step further in the direction of universality:

> How can we do this [follow the poets agains bifurcation]? Whitehead tells us: by not letting nature bifurcate, that is by not letting the primary and secondary qualities go their separate ways. The reception of Whitehead's cosmology over the last century is proof enough that this is not an easy matter. (Latour 2008, 13)[29]

Latour adds, "What is important to remember is that bifurcation is unfair to *both* sides: to the human and social side as well as to the non-human or 'natural' side—a point always missed by phenomenologists" (Latour 2008, 15).[30] Pointing to the understanding of the soul, Latour claims, "It was previously impossible, under bifurcated nature, to ask the question about the monumentality or even objectivity proper to a soul" (Latour 2011, 322). In his fight against the bifurcation of nature, Latour also summons the French sociologist of the early twentieth century Gabriel Tarde (1843–1904), whom he sees as a precursor to a "new sociology." According to Latour, Tarde's thought emphasized above all a far-reaching definition of society. Here, the basic pantheistic thought of an immanent spirit of nature is reflected and recognizable on micro- and macro-levels:

> But this means that everything is a society and that all things are societies. And it is quite remarkable that science, by a logical sequence of its earlier movements, tends to strangely generalize the notion of society. It speaks of cellular societies, why not of atomic societies? Not to mention societies of stars, solar systems. All of the sciences seem fated to become branches of sociology. (Tarde as quoted in Latour 2008, 15)

In another related sense, we can see how Latour's relativist network diverges from Heidegger's absolutist idea of death, and how his network generates a new way of looking at death. Through the network, humans and nonhumans are closely connected to each other. Latour does not merely emphasize the symmetry between humans and nonhumans: there is *a priori* no possibility for making such a distinction. Seen in this context and in view of the underlying understanding of death, this means that humans always leave traces of their existence in the network after death, and that these traces consist of their traits, energies, or characteristics. The traces of their existence, depending on the context, can even represent the humans in all relevant aspects and completely replace them. Thus, humans continue living (and not merely figuratively through their family or friends), and, even more, they continue to exist in the things of the environment they left behind. Humans leave imprints on the things which are closely connected to them in the collective, such as in the books they wrote, the atmosphere of their houses, in artworks, or in entirely new artifacts, such as Steve Jobs in relation to Apple products or Picasso in relation to "Guernica" (1937).[31] But it is important to note that continuing to live—this continuity in the world—is not to be understood only symbolically but is concrete, material, and capable of effecting change. In other words, every actor changes the world; that is, once a human is born, the world will never be as it was before. Or as Latour claims in *AIME*, "In all the sermons about the beyond, about the 'afterlife,' there must be something that can sometimes ring true" (Latour 2013, 305). Latour adds programmatically, "To be or not to be, this is no longer the question! Experience, at last; immanence, especially" (Latour 2013, 178). This exclamation captures Latour's denial or rather transformation of death.

The above interpretation of Latour's thought is not only based on the main tenets of ANT and *AIME*, but also includes a significant thought that Latour first published in 1995 and once again fifteen years later in the book *On the Modern Cult of the Factish Gods* (2010). In this text, Latour critiques the modern separation of facts and fetishes, that is, the separation of magical artifacts and scientific objects, as well as constructivism and realism. Latour sees not only in etymology a clear, subversive relation between the areas. Under the subheading "How the Moderns Struggle—and Fail—to Distinguish Facts from Fetishes," Latour examines the modern critical thinker who is fixated on denouncing the fetishes of pre-modernity. Latour argues that the modern anti-fetish thinker overlooks how he replaces artificial objects with other artificial objects, causing the (premodern) fetish to remain intact:

> To state it bluntly, the critical thinker will put everything he does not believe in on the list of fairy-objects—religion, of course, but also popular culture, fashion, superstitions, mass media, ideology, and so on—and he will put everything in which he firmly believes on the list of case-objects: economics; sociology; linguistics; genetics; geography; neuroscience; mechanics; and so on. (Latour 2010, 14)

Here, we must note not only the new view of sociology but also how Latour describes the cult of sacral modern science, which is distinguished by its Nobel Prize award ceremony in the halls of the Royal Swedish Academy of Sciences.

As Latour deconstructs the separation of magical objects and science, or superstition and fact, he also opens the door to *premodern* pantheisms and cosmologies, which awakens the dead and brings them back to life. In pantheistic thought, death can be seen differently, as something which does not suggest the need for transcendence. Or as Latour writes:

> To use the language of philosophers, no one has ever been able to distinguish between immanence and transcendence. But this stubborn refusal to choose always shows up, we now understand, as a simple practice, as something that can never be spoken or theorized, even if the "actors-themselves" keep on saying it and describing it in luxurious detail. (Latour 2010, 29)

When humans believe they are being talked to by the dead, this is not nothing and must, according to Latour, be taken seriously as transcendence in immanence. There is no consistent criterion that allows us to ignore such a contact as superstition. In Latour's claim that transcendence cannot be distinguished from immanence, we find yet another analogy to the pantheistic understanding of the world.

To summarize Latour's remarkable perspective on death, we must note two different, diverging strategies. On the one hand, there is no absolute death in Latour's thought, as individuality is eliminated by the network from the very beginning, which is why it is meaningless to speak of a fundamental separation of life and death or of an original and a copy, which both require individuality. On the other hand, all humans and nonhumans leave traces behind in the network, which serve as evidence of their existence and through which they continue to live.

TOWARD A RELIGION OF HUMANS AND NONHUMANS

As demonstrated in the preceding interpretations, Latour unfolds a series of concepts and ideas which, when viewed altogether, allow for a presentiment of a pantheistic religion. To understand these thoughts more clearly, it is helpful to emphasize that the concept of religion comes from Latin, *religare*, and means "to connect." In this context, we may understand religion as a connection between this world and the afterworld—between the factical and the fantastical, "Religion does not have to be 'accounted for' by social forces because in its very definition—indeed, in its very name—it links together entities which are not part of the social order" (Latour 2005, 7). Religion is concerned precisely with what Latour attempts to capture and describe with the concept "factical" (Latour's *terminus technicus* for the connection between "fetish" and "fact"), namely, a hybrid of the sensible and the supersensible. This interpretation is elaborated in *AIME* in light of Latour's theology and new perspective of separation:

> If there is an error to avoid in speaking "about God," it lies in separating continuity from discontinuity, repetition from difference, tradition from invention, subsistence from its renewal, monotheism from polytheism, transcendence from holy immanence. And it is indeed that fatal error that religions were trying to avoid in sketching out, quite explicitly, by successive innovations, the admirable series, yes, let's venture the word, this *network* of transformations: "Holy nation," "God," "Son," "Spirit," "Church," this chain of renewals and wild inventions through which only the continuity of a message could be retained without any content but the reprise itself. (Latour 2013, 315)

And in another context and in line with his Catholic education, he adds to his description:

> I am surprised that the speech condition of religion has disappeared. This is because, I think, it has been loaded with rationalism that makes it completely unable to get at the right tone with which it should be articulated. I am trying to find a way to speak about religion without using the idiom of rationality [. . .] But I know this is even more tricky and dangerous than speaking about science. (Latour 2003, 25)

At the same time, it is crucial to understand that the basic meaning of religion, "to connect," is the synonym of a basic concept of ANT, namely "to associate." The concept of associating is even given a main role in Latour's fundamental reform of sociology: "If *sociology* were (as its name suggests) the science of *associations* rather than the science of the social to which it was reduced in the nineteenth century, then perhaps we would be happy to call ourselves 'sociologists'" (Latour 1993, 205). In *Reassembling the Social: An Introduction to Actor-Network-Theory,* Latour adds, "This is the reason why I am going to define the social not as a special domain, a specific realm, or a particular sort of thing, but only as a very peculiar movement of re-association and reassembling" (Latour 2005a, 7). For Latour, sociology should study and interpret the association of all kinds of heterogeneous elements and extend to—in accordance with what has been mentioned—the realm of humans.

> Even though most social scientists would prefer to call "social" a homogenous thing, it's perfectly acceptable to designate by the same word a trail of *associations* between heterogeneous elements. Since in both cases the word retains the same origin—from the Latin root *socius*—it is possible to remain faithful to the original institutions of the social sciences by redefining sociology not as the "science of the social," but as the *tracing of associations*. In this meaning of the adjective, social does not designate a thing among other things, like a black sheep among other white sheep, but *a type of connection* between things that are not themselves social. At first, this definition seems absurd since it risks diluting sociology to mean any type of aggregate from chemical bonds to legal ties, from atomic forces to corporate bodies, from physiological to political assemblies. (Latour 2005a, 5)

The new sociology that Latour introduces attempts to connect heterogeneous actors in the same network and appears to be tailor-made for the study of the associations of religion. Against the background of Latour's personal interest in religion, its absence is glaring in the passage above. Later in Latour's outlines of a new sociology, he writes,

> For example, if an informant says that she lives "in a God ordained world," this statement is not really different from that of another informant who claims he is "dominated by market forces," since both of these terms—"God" and "market"—are mere "expressions" of the *same* social world. An association with God is not substitutable by any other association, it is utterly specific and cannot be reconciled with another one made up of market forces. (Latour 2005a, 35f)

Latour references directly the association with God and aims to characterize it as a special connection. To understand the foundation of this judgment, we must consider that for Latour there is no difference between transcendence and immanence. This means that the association with God is not unique because it is based on a relation to a transcendental being but because it is based on, like every other worldview, a unique network of associations. The association with God is for Latour no different than associations with other non humans. Moreover, the example of "God and market" is not particularly clear, as liberal economists believe in the market's powers. Referencing Adam Smith, they maintain that the "invisible hand" guides the markets and maintains equilibrium; existing between immanence and transcendence, like God, the invisible hand can even be interpreted as a divine power.[32]

HEIDEGGER AND LATOUR ON WORLD, LIFE, AND DEATH

As we have seen above, Heidegger's and Latour's ways of thinking are sometimes close to each other and at other times in sharp contrast.

First, in both Heidegger and Latour, humans are inseparable from the surrounding world. Both thinkers understand the world as a necessary precondition for every human action. Heidegger attempts to articulate this view of the relational mode of being with the concept of being-in-the-world, which is not only connected etymologically with the meaning of human beings, but also to the fundamental intentionality of Dasein expressed in "Being-in." That is, humans are in their Being always directed toward the world. In this sense human life is always lived in the context of other humans, everyday equipment, and natural things. Similarly, Latour emphasizes above all how humans are parts of networks, which encompass organizations, other humans, and technologies. This collective of humans and nonhumans defines the mode of being of humans—we humans are always already parts of far-reaching networks that shape and limit our mode of operation. Similar to Heidegger's formula of being-in-the-world, Latour also uses hyphens in his concept of the "Actor-Network-Theory" in order to emphasize the inner cohesiveness of actors and networks.

Secondly, Heidegger and Latour diverge when dealing with the meaning and the end of the relatedness of humans to the world. Based on Heidegger's thought, we can see that in Latour's descriptions of human functioning and acting, he makes the mistake of not looking behind the life of "the they." Humans may be complex objects for Latour, but they are treated as *black box* cyborgs; that is, Latour is only interested in their function within the network; as objects they can be replaced by other humans and, according to the symmetry principle, by nonhumans as well. Individuality for Latour is only for a limited time and never to be seen as absolute. In his approach, the self-reflection of the actors plays no significant role. Humans are, in his interpretation of ANT, to be understood as actors: they perform their roles, which are never defined by them:

> An "actor" in the hyphenated expression actor-network is not the source of an action but the moving target of a vast array of entities swarming toward it. To retrieve its multiplicity, the simplest solution is to reactivate the metaphors implied in the word actor that I have used so far as an unproblematic placeholder. It is not by accident that this expression, like that of "person," comes from the stage. Far from indicating a pure and unproblematic source of action, they both lead to puzzles as old as the institution of theater itself [. . .] If we accept to unfold the metaphor, the very word actor directs our attention to a complete dislocation of the action, warning us that it is not a coherent, controlled, well-rounded, and clean-edged affair. (Latour 2005a, 45f)

Latour emphasizes that through the optic of ANT, a "complete dislocation of the action" occurs—away from the humans. In Latour's network, there is no room for authenticity, mineness, or autonomy of the individual components—the ruling factors are heteronomy, peer pressure, and the power of the collective. There is in Latour no difference between the passive adoption of collective action-rules, which characterizes the life of the they, and the active, reflective determination which in Heidegger is connected with authenticity.[33]

Latour is interested in humans only insofar as they are acting, that is, he examines the active, innovative, and curious humans. In Latour, curiosity is even an outstanding mode of life, as it ensures the generation of many new associations in the network. Curiosity marginalizes reflective contemplation (which negates the *action* or actors) and thus expands the network. Seen in this way, curiosity is the drive to infinity, which stands in sharp contrast to Heidegger's concept of the anticipation of death.

A decisive objection to Latour, which Heidegger would agree with, is that Latour fails to explain where the drive to build new associations comes from. At the same time, Latour must explain how life is experienced and understood by actors if it is not in contrast to death—a life in which one's own death has become insignificant or indifferent. This notion of life is so fleeting that it is not possible to identify whether a life has come to an end, much less who or what leads it. Such a life negates the concept of Dasein, as it negates the existence of the present in favor of a being-here-and-there (an absentminded state) in an infinite network. Life that represses and denies absolute death appears to resemble the factical life of humans, but it also robs this life of the possibility of recognizing the sentence to death as an exceptional mode of being. In this sense, Latour's actors *have never been human* and resemble science fiction cyborgs. And Latour's reflections in *AIME* does not change this. It is thus ironic that *An Inquiry into Modes of Existence* also speaks of existence, as existence in Heidegger's sense is precluded from the life of Latour's actors and reserved for the mortals.

Thirdly, is death an end in the never-ending association of networks? Based on previous considerations, this question requires a twofold answer. In Heidegger, the question can be answered affirmatively, since for Heidegger, death is the end of every relation to the world. In this view, he is an absolutist. No one can replace another in death, which is why human life—in the face of death—makes humans reflect on themselves. Death marks the radical negation of being-in-the-world, of Dasein, and also implies for Heidegger the end of the relation of human beings to what Latour defines as the collective. Heidegger understands death as radical and inevitable, so that humans in anticipation of death can be overwhelmed and may try to flee to different distractions and curiosities. But this only helps temporarily. Self-reflection retrieves humans from the fleeing, or, that is, per definition, one can never completely ignore that one is fleeing. One can never actually flee from death, as all humans will be ultimately caught by death. Part of the existence of Dasein is its sentence to death. For Heidegger, life is not authentic if it is not lived in the consciousness of its opposite, of death. Heidegger is therefore concerned not with a *perfect* inauthentic life and death since he holds this to be impossible. Such a thing would be, perhaps, only possible for a purely functional machine-type or deeply religious person, but for Latour's actors, this seems to be the only possibility.

The question raised above would have to be answered negatively if presented to Latour. Since integrity, individuality, originality, and autonomy within the collective are from the very beginning eliminated, it is meaningless to speak of absolute death. Every part of the collective is replaceable, and the collective for this reason does not depend on the life of a particular person. A collective must not have a predefined end, just as they cannot die, and an organization is not *a priori* delivered over to death. This is also why death forms no ending in Latour's concept of a network. It is one gradual transformation among many, and in *AIME* Latour does not deny a continued existence beyond death. Seen from the perspective of Latour's humans, death posits no end, since their will, life, and dreams are identical to or replaceable with others in the collective to which they belong to. When humans die, the great drama continues unhindered according to the plan of the collective.[34] The price for this kind of immortality is, however, high. Humans are, in fact, transformed in Latour's network into extras in a play—into ants, robbed of the possibilities of authentic existence.

In Heidegger's view, Latour's concept of human life would be the enthronement of inauthenticity, and it would represent a principal obstacle to connecting human life with a concept of freedom. This automated lifeform would also pose a great threat, according to Heidegger, which he warns against in his later texts. In the collective of humans and nonhumans, where each element is replaceable and individuality is forsaken, humans are nothing more than components of a mechanical wheelwork who can be given an important function but are in themselves without significance. In the fundamental symmetry of the collective, there is no reason to distinguish between humans and nonhumans, because there is only more or less complex equipment. Under the rule of Latour's collective, human life would be transformed, and the definitive boundary between the mode of function of technology and the actions of human beings would be erased. There would be accordingly no ethics to protect humans from the force of the collective, as the collective determines the relative value of life. In other words, the collective itself determines what is master and what is slave. Those who refuse to be part of the collective are denounced by Latour as insane and anachronistic, even if it's high time to reflect on the danger of the concept of the collective. In the sense of Heidegger, *AIME* shows no way out of this danger, quite the contrary. As we analyze Latour's work in more detail, we find that there is indeed a collective fleeing from human existence. In the end, Latour's actors *have never been authentically human*; he has confused them with stage players and cyborgs who are ordered to act.

NOTES

1. Cf. Heidegger (2008, 53) and Riis (2018).
2. In this book, I make no systematic distinction between pantheism and animism. It is important to note that both of these ideas emphasize a kind of symmetry between humans and nonhumans and highlight a pervasive steering principle across all beings. However, it is possible to excavate a number of differences between the two concepts. In order to follow the latter interest, cf. Melson, A. and Riis, S. (2015).
3. Cf. Heidegger (2008, 135) and (Tasheva 2001).
4. Cf. Riis (2018).
5. Cf. Heidegger (2008, 236 and 245).
6. Cf. Heidegger (2008, 236).
7. Cf. Tasheva (2001, 163).
8. Cf. Heidegger (2008, 239).
9. For an outstanding description, cf. Jünger (1982).
10. Cf. Heidegger (2008, 238).
11. Cf. Tasheva (2009, 285).

12. Death is Dasein's *ownmost* possibility [. . .] The ownmost possibility is *non-relational*. Anticipation allows Dasein to understand that potentiality-for-being in which

its ownmost Being is an issue, must be taken over by Dasein alone. Death does not just "belong" to one's own Dasein in an undifferentiated way; death *lays claim* to it as an *individual* Dasein. The non-relational character of death, as understood in anticipation, individualizes Dasein down to itself. This individualizing is a way in which the "there" is disclosed for existence. It makes manifest that all Being-alongside the things with which we concern ourselves, and all Being-with Others, will fail us when our ownmost potentiality-for Being is the issue. (Heidegger 2008, 263)

13. Cf. Heidegger (2008, 188 and 240).
14. Cf. Heidegger (2008, 242).
15. Cf. Tasheva (2001, 166). It is also important to understand the analogy between the beginning and end of life in Heidegger. Both are significantly shaped by discontinuity. At the end of life, there is death, and at the beginning, there is the phenomenon of thrownness (*Geworfenheit*), that is, according to Heidegger, something which Dasein "is and has to be" (2008, 135).
16. Cf. Heidegger (2008, 134).
17. See also Annette Hilt's examination (2005, 296).
18. Cf. Carr (2008).
19. Cf. Heidegger (2008, 188).
20. Cf. Johannes Weiß (2001, 36), who interprets existential solipsism in view of Heidegger's notion of mineness.
21. In this sense, the foundation of ANT shares many similarities with the basic principles of cybernetics:

We believe that men and other animals are like machines from the scientific standpoint because we believe that the only fruitful methods for the study of human and animal behaviour are the methods applicable to the behaviour of mechanical objects as well. Thus, our main reason for selecting the terms in question was to emphasize that, as objects of scientific enquiry, humans do not differ from machines. (Rosenblueth and Wiener 1950, 328)

Cf. La Mettrie (1912, 148), Peter Galison (1994), and Pickering (2002). Heidegger does not support such an approach and is critical of cybernetics (Riis 2018).
22. Cf. Nitsch (2008, 223).
23. Cf. Harman (2009, 122).
24. Cf. Milgram (1967). Although there are more humans on Earth today than when the theory was developed, the Internet has extended the network of the maority of humans today and thus reaffirming the theory. See also Markoff and Sengupta (2011).
25. Confer also the ANT-inspired understanding of a mummy, in which

a mummified body is precisely—and virtually by definition—a well-constructed and enduring object. What the Egyptians knew so well was that in order for the deceased to subsist, despite the gnawing teeth of time, he or she had to be prepared, embalmed, encapsulated in layers of wrappings and laid to rest in a still environment. In other words: for a body to become a mummified body it had to be carefully and properly made. (Munk and Abrahamsson 2012, 57)

26. Cf. Bennett (2010).

27. Harman writes, "Latour is reluctant to believe that anything substantial could exist outside of all networks" (Harman 2002, 312).

28. Latour states,

> Strangely enough, we have changed time so completely that we have shifted from the time of Time to the time of Simultaneity. Nothing, it seems, accepts to simply reside in the past, and no one feels intimidated any more by the adjectives "irrational," "backward" or "archaic" [. . .] Everything has become contemporary. (Latour 2005b, 40)

29. Cf. Gill (2008, 47).

30. How did we manage to behave as if Nature had "bifurcated" into primary qualities—which, if you remember, are real, material, without value and goals and only known through totally unknown conduits—and secondary qualities which are nothing but "psychic additions" projected by the human mind onto a meaningless world of pure matter and which have no external reality although they carry goals and values. How did we succeed in having the whole of philosophy reduced to a choice between two meaninglessnesses: the real but meaningless matter and the meaningful but unreal symbol? (Latour 2008, 36)

31. Cf. Guynn (2012). In this context, a statement made by the icon of street art, Banksy, is also interesting: "I mean, they say that you die twice. Once when you stop breathing, and again, a bit later, when someone says your name for the last time" (Banksy 2014).

32. Cf. Smith (2010, Book IV).

33. Cf. Riis (2018).

34. Cf. Tasheva (2009).

Chapter 4

Another Beginning

New technologies have been heavily criticized for their negative impact on human beings in the philosophical tradition of the twentieth century. Despite philosophical critique, civil society has enthusiastically created and embraced thousands of new breakthrough technologies.[1] This more or less pronounced confrontation between philosophical theory and human practice is still largely unresolved.

> How can we relate ourselves to technology in a way that not only resists its devastation but also gives it a positive role in our lives? This is an extremely difficult question to which no one has yet given an adequate response, but it is perhaps the question for our generation. (Dreyfus and Spinosa 1995, 159)

To understand the tension and reconcile these two spheres, this chapter will show how Heidegger and Latour may be read in support of a poetic practice of technology.

The disconnect between Heidegger and Latour diverts our attention away from the number of shared characteristics that show a high degree of continuity between them. In this chapter, I will argue that Latour can be read as a thinker who works out a constructive and viable path between technology pessimism and optimism, which Heidegger was also trying to uncover and describe (Heidegger 1977, 311f). Artifacts may become the gathering places capable of connecting these two thinkers, who, in many ways, appear to be counterparts. In other words, Latour may be read as a thinker who, inspired by Heidegger, offers concrete suggestions for a complex and edifying understanding of the relationships between humans and technologies. Latour moves beyond the inquiries of artifacts in the shape of technology pessimism and optimism and prepares and unfolds a different point of departure for understanding and assessing technologies.

My claim is that Latour builds upon Heidegger's understanding of technology. I shall argue that Heidegger's legacy is essential to Latour insofar as he unfolds this understanding of technology parallel to Heidegger. The most immediate problem regarding the association between the respective understandings of technologies has to do with the fact that research on Heidegger's understanding of technology is primarily based on the text "The Question Concerning Technology" (Heidegger 1977a). This starting point is neither inevitable nor harmless: Heidegger engages with technologies in several texts, and "The Question Concerning Technology" mainly focuses on the profound dangers related to modern technology.

As indicated in the title of this chapter, "Another Beginning" refers to the inspiration that Latour finds in Heidegger, which allows him to develop an alternative to both technology pessimism and technology optimism. But the title also indicates that I, along with Latour, aim to develop a more constructive understanding of technology than the one forwarded in "The Question Concerning Technology." Through a reinterpretation of Heidegger's lecture "The Thing" (Heidegger 2001), a different approach to technologies is taken up, resulting in a new understanding of how to interpret technologies. "The Thing" complements Heidegger's other critical reflections on modern technologies with a more balanced and more enriching interpretation of artifacts as a place of disclosure, which Heidegger seeks to capture with his renowned yet convoluted concept of "the fourfold." Following this idea, I argue that Heidegger's notion of the fourfold provides inspiration and guidance for a more constructive interaction with modern technologies. In particular, I endeavor to show this using the example of a modern car and airplane. The notion of the fourfold also illuminates the contours of what we shall come to see as the Heideggerian "saving power" of artifacts. More precisely, the fourfold manifests basic hermeneutic and ontological insights that may *save us* from the one-dimensional, negative interactions with modern technologies that lead to the advancement of "enframing" and the stockpiling of resources. My intention here is to demystify and reformulate the concept of the fourfold and show its proper, wide-ranging scope. The fourfold creates space and opportunities for thinking by revealing a poetic opening to modern technologies.

BEYOND READY-TO-HANDEDNESS AND PRESENT-AT-HANDEDNESS

A few years before Heidegger inquired into the essence of technology in "The Question Concerning Technology," he gave a fascinating lecture on the particular being of things entitled "The Thing." This lesser-known lecture brought modern technology into play in a different and ultimately more edifying way than both the early seminal work *Being and Time* and the so-called "technology lecture." Heidegger begins "The Thing" by stating that a number of new technological innovations have reduced the distances in space and time. The modern airplane allows us to suddenly decrease distances that used to take weeks or months to traverse to only a few hours or days (Heidegger 2001, 165). And the radio informs us about events almost as they are taking place—events that would have taken several years to spread before this innovation (Heidegger 2001, 165). As the distances in time and space continue to be reduced today, not least due to the Internet, this phenomenon is catalyzed and thus seems even more urgent to address.[2]

Following his initial considerations, Heidegger noted that the new technological innovations did not, paradoxically, create more *nearness* despite this revolution of distance. In the rush to become master of space and time, *modern humans* have failed to get truly near to the various objects and places in the world. Instead of redefining the dimensions of the world, dimensionality seems to have disintegrated in the world of the moderns. According to Heidegger's critique, the rapid overcoming of space and time even blocks us from engaging in *near* relationships. Presaging his analysis, Heidegger poses a coherent set of questions in an attempt to get closer to the meaning of nearness:

> What is nearness if it fails to come about despite the reduction of the longest distances to the shortest intervals? What is nearness if it is even repelled by the restless abolition of distances? What is nearness if, along with its failure to appear, remoteness also remains absent? [. . .] What is this uniformity in which everything is neither far nor near—is, as it were, without distance? (Heidegger 2001, 165f)

The dissolution of the dimensions of space and time for the sake of a pervasive sameness is a modern phenomenon and is, in Heidegger's optics, dangerous and requires careful attention. When the dimensions dissolve, humans cannot find their way and become homeless in a profound and disturbing sense. It is also in this context that Heidegger makes one of his rather infamous references to the atomic bomb. Heidegger delivered his lecture "The Thing" precisely in the period after the Second World War, when there was a widespread fear of the atomic bomb. The point of Heidegger's example, however, is to demonstrate that the dangers of the atomic bomb are insubstantial compared to the dangers of the far and deep-reaching revolution of space and time stemming from the abolition of space and time and the repeal of nearness. Heidegger asserts:

Everything gets lumped together into uniform distancelessness. How? Is not this merging of everything into the distanceless more un-earthely than everything bursting apart? Man stares at what the explosion of the atom bomb could bring with it. He does not see that the atom bomb and its explosion are the mere final emission of what has long since taken place, has already happened. (Heidegger 2001, 166)

In this passage, Heidegger suggests that there is a fundamental change of all beings taking place—a change which has gone relatively unnoticed until now. Or stated differently, the technological innovations leading up to this change were even applauded; however, the flip side has never really been identified, understood, or dealt with. The atomic bomb represents a different situation since the negative consequences were obvious after the Second World War.[3] If new technologies continue to be developed, unhampered by war or the explosion of more atomic bombs, we should not feel safe. On the contrary, we should fear the uncanny technological dissolution of space and time that continues unabated (Heidegger 1977).

In an attempt to reclaim and rehabilitate space and time as fundamental dimensions in the world and as preconditions for humans to experience nearness, Heidegger takes a detour in his lecture. As unfolded above, nearness is hard to experience and appreciate at first glance, as it appears to be dissolved in the modern high-tech world. Heidegger therefore asks if anything can be experienced as near at all. In response, Heidegger claims that we experience "the things" as being nearby: things such as a jug, a chair, or a picture, or even a tree or a mountain in proximity (Heidegger 2001, 166). Before Heidegger is able to qualify the things further, he is forced to clear some misunderstandings out of the way, including the (mis)understandings of how things are manifested in the natural sciences, where such a thing as a jug is grasped quite abstractly, which hinders nearness.[4] Heidegger maintains that the sciences miss the everyday experience with basic things such as jugs:

When we fill the jug, the pouring that fills it flows into the empty jug. The emptiness, the void, is what does the vessel's holding. When we fill the jug, the pouring that fills it flows into the empty jug. The empty space, this nothing of the jug, is what the jug is as the holding vessel [. . .] And yet is the jug really empty? Physical science assures us that the jug is filled with everything that goes to make up the air's mixture. We allowed ourselves to be misled by a *semipoetic* way of looking at things when we pointed to the void of the jug in order to define its acting as a container. But as soon as we agree to study the actual jug scientifically, in regard to its reality, the facts turn out differently. When we pour wine into the jug, the air that already fills the jug is simply displaced by a liquid. Considered scientifically, to fill a jug means to exchange one filling for another. These statements of physics are correct. By means of them, science represents something real, by which it is objectively controlled. But—is this reality the jug? No. Science always encounters only what *its* kind of representation has admitted beforehand as an object possible for science [. . .] Science makes the jug-thing into a non-entity in not permitting things to be the standard for what is real. (Heidegger 1997, 169f; emphasis added)

How the jug is integrated into everyday social practices fundamentally differs from the scientific view of it. In our everyday practical dealings with the jug, we take into account whether it is designed to hold wine, coffee, or other drinks, and keep in mind why it is being used, for example, it may be a festive occasion or an everyday event. These phenomenological considerations connected to the lifeworld of the users are ignored by the general abstract perspective of the natural sciences, according to Heidegger.

His criticism of the natural sciences is also apparent in the concept of nearness, which can be distinguished from his concept of "present-at-handedness." Heidegger developed present-at-handedness in relation to the natural sciences in *Being and Time*. It is the relationship between humans and objects and is characterized by equal measures of distance and reduction:

And only when innerworldly beings can be encountered at all does the possibility exist of making what is merely objectively present [present-at-hand/ Vorhanden] accessible in the field of these beings. On the basis of their merely objective presence these beings can be determined mathematically in "functional concepts" with regard to their "properties." (Heidegger 1996, 83)

Conversely, the nearness characterizing the relationship with the jug is also not exhausted in Heidegger's concept of "ready-to-handedness," which he wrote more than twenty years earlier—not even if we take the cup in *hand* when using it. As Heidegger slowly advances an understanding of nearness to the things around us, including the jug, it becomes clear that the nearness presented in "The Thing" has a different character. It is more associative and poetic, and it is not expressed in the completely transparent relationship of ready-to-handedness. In *Being and Time*, Heidegger writes about ready-to-handedness:

We shall call the useful thing's kind of being in which it reveals itself by itself *handiness*. It is only because useful things have this "being-in-themselves," and do not merely occur, that they are handy in the broadest sense and are at our disposal. [. . .] When we just look at things "theoretically," we lack an understanding of handiness. But association which makes use of things is not blind, it has its own way of seeing which guides our operations and gives them their specific thingly quality. (Heidegger 1996, 65)

The jug is, of course, close to our hands, and based on *Being and Time*, our relation to the jug could be assessed according to its ready-to-handedness.

But nearness reveals another significant dimension of human interaction with beings, one that is distinct from ready-to-handedness and present-at-handedness.[5]

When the jug is experienced as *near*, it is neither overlooked nor does it disappear in an abstract gaze or its concrete use. If we simply embed the jug in our practice, we will overlook the jug (i.e., in the context of use, in which the jug is used as a means to quench thirst, nearness will not automatically emerge). A jug that is experienced and understood as *near* is characterized by its ability to gather and hold wine and water; it is the jug that bestows the tasting of the wine and is able to give the wine its provision. In view of the jug's full context, which it enters into and reflects, Heidegger writes poetically and associatively about the space opened up by the jug:

> The spring stays on in the water of the gift. In the spring the rock dwells, and in the rock dwells the dark slumber of the earth, which receives the rain and dew of the sky. In the water of the spring dwells the marriage of sky and earth. It stays in the wine given by the fruit of the vine, the fruit in which the earth's nourishment and the sky's sun are betrothed to one another. In the gift of water, in the gift of wine, sky and earth dwell. (Heidegger 2001, 172)

Heidegger unfolds the bestowing character of the jug. In this sense, it is a modification of a sacrifice. It is analogous to the archetypal religious sacrifices that are capable of connecting mortals with the divinities (Heidegger 2001, 172). In other words, as reflected in the quotation above, Heidegger believes that the jug manifests a collection of the following four elements: earth, sky, divine, and mortal. Heidegger entitles the unity of these four elements "the fourfold" (*das Geviert*) (Heidegger 2001, 179).

The relationship that we are dealing with here is thus very different from the relationships described as ready-to-handedness and present-at-handedness in *Being and Time*, which are both connected to the use and scientific analysis of equipment (*Zeug*). In *Being and Time*, equipment is part of a practical context that Heidegger calls the "totality of equipment" (*die Zeugganzheit*). The practical context described here with down-to-earth, pragmatic terms is quite distinct from the relationship that the jug bestows access to in "The Thing":

> Strictly speaking, there "is" no such thing as *a* useful thing [equipment/*Zeug*]. There always belongs to the being of a useful thing a totality of useful things in which this useful thing can be what it is. A useful thing is essentially "something in order to . . ..". The different kinds of "in order to" such as serviceability, helpfulness, usability, handiness, constitute a totality of useful things. The structure of "in order to" contains a *reference* of something to something. [. . .] In accordance with their character of being usable material, useful things always are in terms of their belonging to other useful things: writing materials, pen,

ink, paper, desk blotter, table, lamp, furniture, windows, doors, room. These "things" never show themselves initially by themselves, in order to then fill out a room as a sum of real things. What we encounter as nearest to us, although we do not grasp it thematically, is the room, not as what is "between the four walls" in a geometrical, special sense, but rather as material for living. On the basis of the latter we find "accommodations," and in accommodations the actual "individual" useful thing. A totality of useful things is always already discovered *before* the individual useful thing. (Heidegger 1996, 64)

Heidegger's description of equipment and things in *Being and Time* shares a number of characteristics with his depiction of them in "The Thing": they are both embedded in a practice and they are part of and reveal a broader context. The difference has to do with us, as the users of equipment. Because we often overlook the larger context—the network in which equipment and things are embedded—we have not cultivated the sense of care necessary to see, understand, or appreciate the relationships that are gathered and reflected in the jug, that is, it is not experienced as *near*. In other words, it appears as if Heidegger in *Being and Time* views the thing as closer to poetic thinking than the equipment—in spite of terminological overlaps. Based on Heidegger's description, it is apparent that there are two different contexts: "total relevance"[6] (*Bewandtnisganzheit*), which is quite practical and ordinary, and the relation of "the fourfold," which discloses the poetic dimensions of the meaning of things. Stated differently, for Heidegger *the thing* reveals a greater interconnected tangible and intangible network of meanings that is covered up by the ready-to-handedness of the equipment in use.[7]

As Heidegger writes in the passage above, the *room* is revealed before the equipment. With the concept of the fourfold, Heidegger analogously tries to identify the dimensions in the overarching whole, *the world*, that stipulate the creation of meaning in the first place. In both contexts, a relationship exists between the parts and the whole. This reciprocal relationship as such is reflected in the connection between the thing and the fourfold. Taken to the extreme, there is a thought-provoking relationship between the jug and the divine that creates associations and gathers the world we inhabit. Things reveal the fourfold and make its dimensions tangible. This relationship has no scientific, logical, or necessary character; conversely, it is not impossible but poetic and associative in nature and stems from the richness of lifeworld experiences (Heidegger 2001, 170).[8] If we revisit the passage from the text "Poetically Man Dwells" (Heidegger 2001b) in which Heidegger emphasizes the primacy of the dimensions over the room, we may get a better understanding of the reciprocity between things and the fourfold and learn to see how the dimensions of the fourfold create a poetic space for thinking:

Nor is the dimension a stretch of space as ordinarily understood; for everything spatial, as something for which space is made, is already in need of the dimension, that is, that into which it is admitted. [. . .] Man does not undertake this spanning just now and then; rather, man is man at all only in such spanning. (Heidegger 2001b, 218)

It is also important to note that, in Heidegger's view, the thing has more autonomy than the equipment, which is constantly being used for other tasks (i.e., it points away from itself) and is essentially overlooked; it is precisely the thing that gathers attention and reveals relations. This distinction may seem insignificant, but it is what separates the restless movement of the elements in the network of "total relevance" (*Bewandtnisganzheit*) from the dwelling attention attached to things.

Having unfolded the jug's mode of being as a gathering of the fourfold, Heidegger declares that this unifying characteristic is precisely the hallmark of things. As he points out, the etymology of the thing is "assembly" in accordance with the Germanic concept of "the thing" (Heidegger 2001):

The jug's presencing is the pure, giving gathering of the onefold fourfold into a single time-space, a single stay. The jug is the jug as a thing. But how does the thing presence? The thing things. Thinging gathers. Appropriating the fourfold, it gathers the fourfold's stay, its while, into something that stays for a while: into this thing, that thing. (Heidegger 2001, 174)

The jug paradigmatically expresses the characteristics of the thing as it readily reveals the unity of the fourfold.

As we shall see in further detail later in the chapter, Heidegger's understanding of the thing may also be seen as a critique of Kant's Copernican revolution, since Heidegger's focus on things stands in direct contrast to Kant's abstract, pure categories of perception and apperception that subordinate the things under the human faculty of knowledge.[9] Unlike Kant, Heidegger argues that it is actually humans who depend on the things (*bedingte Wesen*), "In the strict sense of the German word *bedingt*, we are the be-thinged, the conditioned ones" (Heidegger 2001, 181). This means that the things bestow measure and meaning on humans. According to this interpretation, space is neither a product of our imagination nor simply an empty container that can be filled up with things. Just the opposite, Heidegger first focuses on the things, which gather, create nearness, and retain remoteness as such: "The Thing is not 'in' nearness, 'in' proximity, as if nearness were a container. Nearness is at work in bringing near, as the thinging of the thing" (Heidegger 2001, 178). The things create the fundamental dimensions in our world and pave the way for nearness.

We have now gained insight into how Heidegger understands and thinks about the nearness that things create by gathering the fourfold. Humans make up a key dimension of the fourfold, but they are just one dimension among four. The experience of nearness, as well as the relative (and humble) position of humans in the world, in opposition to Kant's epistemology, are important elements of Heidegger's interpretation of the nature of things. These are also some of the elements that Latour explicitly affirms, which will be examined in the final section concerning the possibilities of creating nearness in relation to modern artifacts and reinterpreting the fourfold.

LATOUR'S ARK

Latour has repeatedly denounced Heidegger's thinking while at the same time recognizing and working explicitly with Heidegger's archaic as well as pioneering concept of "the thing," and he is also attracted to the concept of the fourfold. Like Heidegger, Latour stresses the ability of things to gather attention and generate interest; through the prism of things, the world becomes more interconnected and truly visible.

It is in the context of the Heidegger-inspired concept of the thing that Latour's ambitious anthology and its accompanying exhibition of artifacts should be understood, both under the name *Making Things Public* (2005a). In this anthology, Latour brings together a number of leading international scholars to express their thoughts about all kinds of *things* in order to show how these particular things gather our attention, are meaningful in their own way, and are essential to human (co)existence in the world. The anthology functions as an advanced exhibition catalog, which manifests why things are not only to be considered simple, immutable matters of fact, but rather as important and relevant matters of concern.

In the introduction to the extensive anthology, Latour explicitly references Heidegger, yet, ironically, tries to downplay his influence: "As every reader of Heidegger knows, or as every glance at an English dictionary under the heading 'Thing' will certify, the old word 'Thing' or 'Ding' designated originally a certain type of archaic assembly" (Latour 2005a, 22). After this general observation, Latour expresses his connection to Heidegger more pointedly, but also explains how his thought differs:

> *Gatherings* is the translation that Heidegger used, to talk about those Things, those sites able to assemble mortals and gods, humans and non-humans. There is more than a little irony in extending this meaning to what Heidegger and his followers loved to hate, namely science, technology, commerce, industry and popular culture. (Latour 2005a, 23)

Through "the fabulous power of their etymology"[10] (Latour 2005a, 119), originally pointed out by Heidegger in "The Thing," Latour confirms half a century later than Heidegger the amazing character of things. But, according to Latour, the notion of things should not be restricted to "stones, rugs, mugs and hammers" (Latour 2005a, 114) but include modern technological products.[11] While Latour explicitly affirms the legacy of Heidegger, he also shuns the narrowing of the concept of things that Heidegger is in favor of, which is evident in his taking a jug as the focal point of his consideration of things. In other words, Latour strives, ironically, to achieve a more *universal* concept of the thing than Heidegger. In contrast to Heidegger, who at first glance sees the unifying character only in archaic things, Latour sees this unifying character in *all* things.

Latour's vision is encompassed by the notion of the thing: it is the answer to how the world avoids atomization, how politics may regain interest and attention, and the type of research that should be carried out. In programmatic terms from Latour's introduction to *Making Things Public*, "There might be no continuity, no coherence in our opinions, but there is a hidden continuity and a hidden coherence in what we are attached to" (Latour 2005a, 15). This claim is underlined in his description of the proposed intention of the book and exhibition:

> Each object may also offer new ways of achieving closure without having to agree on much else. In other words, objects—taken as so many issues—bind all of us in ways that map out public space profoundly different from what is usually recognized under the label of "the political." It is this space, this hidden geography that we wish to explore through this catalogue and exhibition.[12] (Latour 2005a, 15)

The importance of the hidden political geography is highlighted by the extent of the anthology and the range of renowned scholars who contribute to the book, which reveal vital aspects of a variety of *things*, that is, their history, character, performance, and so on.

Both Latour and Heidegger are aware that we humans and our institutions often ignore and override the nature of things, which Latour associates with a special danger. There is an important need for us to realize how things are closely connected to us, and how we must begin to organize ourselves in close proximity to them before we can examine our relationship with them. These things, these potential gathering places, these "small parliaments," as Latour describes them, are crucial to the world's continued existence (Latour 2005a, 31f). We have been, according to Latour, led astray. We have lost our sense of direction, and he warns of impending disaster if we do not respect the significance and fundamental importance of things. Stated differently, Latour hopes that *Making Things Public* will contribute to a comprehensive reflection on things and inform humans about how they can gather materials to aid in their survival and become sustainable in the broadest sense. In this sense, the exposition itself may be seen as a paradigmatic assembly of things, "In this show, we hope visitors will shop for the materials that might be needed later for them *to build this new Noah's Ark*: the Parliament of Things. Don't you hear the rain pouring relentlessly already?" (Latour 2005a, 34; emphasis added).[13]

In the anthology's introduction, Latour paints a bleak picture of the current state of affairs and the occasion for the exhibition. Latour's introduction thus shares a number of features with Heidegger's preliminary observations in "The Thing." The rain, which Latour interprets as a bad omen in the quote above, describes the current crisis, which manifests itself in a *relentless* way (i.e., it is intense and shows no mercy). In other words, Latour sees clear signs that we are living in dangerous times that call for a revolution in our understanding of the world—a kind of revolution that is analogous to that which Heidegger considers necessary. It is crucial that we begin to take things much more seriously and put them at the center of our political labor, that is, we should see them as the nodes in our collective arrangements and associations. If we fail to do so, we will become *homeless* in a fundamental sense, as the pending disaster will materialize and make it impossible to find a place *to gather*.[14]

Latour understands politics as a kind of thinking and acting that is concerned with *things*. It can lead the way out of the current crisis of collective organizations, which repeatedly fail because they do not pay sufficient attention to things. To take things seriously means to regard and treat them not only as matters-of-fact but as matters-of-concern. Things are essential for all life on earth. A thing is something we have to worry and care about:

> For too long, objects have been wrongly portrayed as matters-of-fact. This is unfair to them, unfair to science, unfair to objectivity, unfair to experience. They are much more interesting, variegated, uncertain, complicated, far reaching, heterogeneous, risky, historical, local, material and networky than the pathetic version offered for too long by philosophers. (Latour 2005a, 19f)

Yet, philosophers are not all the same, and Heidegger is, as we have seen above, an exception to the philosophers Latour has in mind.

From Latour's statement, it is now possible to understand how the concept of *making things public* is both the objective and the epitome of Latour's concept of the political. Politics consists of bringing people together (in a kind of *thing*) and discussing things as matters-of-concern. These two movements are two sides of the same coin, since we can only gather if a kind of arena has been created for this purpose (an assembly) and if there is something at stake that we really care about (matters-of-concern).

Latour believes this conceptualization of things and politics is nowadays inseparable from modern technologies and sciences. If there is something that really relates to and influences our lives today, and that may serve as platforms for discussions and debate, it is technologies and scientific objects and practices. In this context, the political challenge is to learn how to explore and discuss the many new things, or artifacts, that technology and science produce beyond the limits of their most basic manifestations. We need to understand in which way an artifact is, in fact, "interesting, variegated, uncertain, complicated, far reaching, heterogeneous, risky, historical, local, material and networky" (Latour 2005a 19f). Because new technological artifacts are key players in politics, political thought should be able to show their far-reaching consequences.

Ontology and political philosophy have failed to help us understand the complicated and "networky" character of things:

> It's not unfair to say that political philosophy has often been the victim of a strong object-avoidance tendency. From Hobbes to Rawls, from Rousseau to Habermas, many procedures have been devised to assemble the relevant parties, to authorize them to contract, to check their degree of representativity, to discover the ideal speech conditions, to detect the legitimate closure, to write the good constitution. But when it comes down to what is at issue, namely the object of concern that brings them together, not a word is uttered. In a strange way, political science is mute just at the moment when the objects of concern should be brought in and made to speak up loudly.[15] (Latour 2005a 16)

A good example of a concrete artifact—a matter-of-concern—with far-reaching consequences and a widespread network with other artifacts is the car, which was invented in the early twentieth century. A closer look at this artifact may show how Latour departs from Heidegger, yet exceeds his exemplifications of things.[16] For many, the car is basically a means to transport individuals from point A to B. This is not wrong, it is even quite *right*, but the car has numerous modes of existence.[17] It is closely linked to a global infrastructure, which has led to the expansion of the freeway system, as well as the production, transportation, refinement, and distribution of oil, gasoline, electricity, and hydrogen. In addition, the car is closely tied to debates about climate change, how Western culture understands itself, and the issue of road safety, as well as notions of freedom, progress, and happiness. All of these elements—alongside countless others—are at stake and should be discussed if we are to understand the car.[18] In this way the car turns out to be a paradigmatic matter-of-concern.

If we are to practice good and sustainable politics in Latour's sense, we have to affirm the inquiry into modern technologies and accept the kind of complexity that fundamentally links people and things. We must cultivate debates about the alliances between people and things: how are both represented together in all sorts of critical contexts; how do humans live and belong among the things; and what about the mediation at play between all parties (i.e., interaction, sharing, dissemination, and negotiation). In Latour's view of the essential task of our thinking and action today, mediation replaces the notion of the evident and complexity replaces immediacy. Latour makes clear references to Heidegger, yet distinguishes himself by not excluding complex technological artifacts from the good (political) company.

TOWARD A HERMENEUTICS OF ALL THINGS

As we have seen, Latour affirms the importance of Heidegger's concept of the thing in his attempt to reformulate a positive vision of how assembly, mediation, and politics can take place and succeed today. Latour believes that we can create a concrete and meaningful understanding of the common issues that challenge us and require our attention if we base it on things. But Latour explicitly distances himself from Heidegger in his attempt to understand things in a broader sense than Heidegger himself suggested. Latour's things are not only "stones, rugs, mugs and hammers" (Latour 2005a, 114) but also more complex (modern) artifacts such as televisions, blogs, cars, shuttles, photographs, and atomic bombs, which are all (re)presented in *Making Things Public* (Latour 2005a). And it is in this sense that Latour's concept of a thing may be described as more universal than Heidegger's.

In this third and final section of the chapter, I will examine how Latour's interpretation and elaboration of Heidegger's concept of things are both compatible and incompatible with Heidegger's thinking concerning things. My claim is that Latour points out an important limitation of Heidegger's concept of things, but that it is possible to reread Heidegger in a way that removes this limitation yet values Heidegger's fundamental insights. In other words, the aim of this section is to present a more comprehensive, complex, and mediated interpretation of Heidegger's thinking concerning things—an interpretation which is inspired by Latour's concept of a thing. As I expand upon the connection to Heidegger's thought, the conception of a thing gains a different philosophical gravity than found in Latour. Since the task is to generate a less archaic and philosophically more vibrant notion of a thing, I will first briefly point to Heidegger's text entitled "Building Dwelling Thinking" (Heidegger 1977f), which was written in the period when he gave the lecture "The Thing" and published in the same book (Heidegger, 1994f).

"Building Dwelling Thinking" has been studied intensively by the two philosophers Hubert Dreyfus and Charles Spinosa, who have jointly written the article "Highway Bridges and Feasts: Heidegger and Borgmann on How to Affirm Technology" (Heidegger 1995). In their article, they focus on an issue analogous to the one being developed here, but their premises and results differ insofar as they do not connect to Latour's insights and give less attention to interpreting and reactualizing the concept of the fourfold.

Dreyfus and Spinosa commence their inquiry by asking a research question, which is followed by a programmatic statement that emphasizes not only the importance of their undertaking but also the significance of the present study as well,

> How can we relate ourselves to technology in a way that not only resists its devastation but also gives it a positive role in our lives? This is an extremely difficult question to which no one has yet given an adequate response, but it is perhaps *the* question for our generation. (Dreyfus and Spinosa 1995, 159)

Dreyfus and Spinosa then, akin to Heidegger, show the dangers stemming from new technologies, but emphasize a key example in "Building Dwelling Thinking" in which Heidegger actually associates his concept of the fourfold with modern technology *par excellence*, namely, an autobahn bridge. With reference to "Building Dwelling Thinking," Dreyfus and Spinosa write:

> But if there were a way that technological devices could thing and thereby gather us, then one could be drawn into a positive relationship with them without becoming a resource engaged in this disaggregation and reaggregation of things and oneself and thereby losing one's nature as a discloser. Precisely in response to this possibility, Heidegger, while still thinking of bridges, overcomes his Black Forest nostalgia and suggests a radical possibility unexplored by Borgmann. In reading Heidegger's list of bridges from various epochs, each of which things inconspicuously "in its own way," no one seems to have noticed the last bridge in the series. After his kitschy remarks on the humble old stone bridge, Heidegger continues: "The highway bridge is tied into the network of long-distance traffic, paced as calculated for maximum yield." Clearly Heidegger is thinking of the postmodern autobahn interchange, in the middle of nowhere, connecting many highways so as to provide easy access to as many destinations as possible. Surely, one might think, Heidegger's point is that such a technological artifact could not possibly thing. Yet Heidegger continues: "Ever differently the bridge escorts the lingering and hastening ways of men to and fro [. . .] The bridge gathers, as a passage that crosses, before the divinities—whether we explicitly think of, and visibly give thanks for, their presence, as in the figure of the saint of the bridge, or whether that divine presence is hidden or even pushed aside."[19] (Dreyfus and Spinosa 1997, 169f)

In light of Dreyfus and Spinosa's research, it is apparent that Heidegger's thinking allows for a modern highway bridge to also gather the fourfold. Based on Heidegger's own concrete example of an edifying interpretation of modern technology, I will now move along a different path than the one Dreyfus and Spinosa embarked upon, since I will connect the example above with Heidegger's final thoughts in his groundbreaking lectures on technology, "The Question Concerning Technology." In the final paragraphs of the technology lecture, Heidegger emphasizes in chorus with the poet Friedrich Hölderlin that humans inhabit the earth poetically , and that one day it will be possible for technology to appear in a new and different way:

> Yet we can be astounded. Before what? *Before this other possibility*: that the frenziedness of technology may entrench itself everywhere to such an extent that someday, throughout everything technological, the essence of technology may unfold essentially in the propriative event of truth. (Heidegger 1977a, 340; emphasis added)

In the lecture, Heidegger does not specifically unfold this crucial passage but mentions its potential. By taking "The Thing" as our point of departure and connecting it with "Building Dwelling Thinking" and "The Question Concerning Technology," I endeavor to show how this potential may be realized, so that utensils, equipment, and artifacts can step out of a sort of one-dimensional shadow sense of being and be recognized possibly as full-blown things in Heidegger's sense.[20]

If we take into consideration Heidegger's own optimism and acknowledgment of another beginning,[21] as well as his example of the modern autobahn bridge, then Heidegger's and Latour's respective understandings of the concept of the thing should be reexamined in order to determine how closely these thinkers are actually linked together. In this reinterpretation, I begin with an interpretation of the four dimensions making up the fourfold; this step allows me to identify the concept's potential and examine whether it can be associated with Latour's insights and enrich our understanding and experience of modern technologies. More precisely, I claim that the fourfold generates basic hermeneutical insights that overcome a one-dimensional interaction with modern technology. My intent is to give the concept of the fourfold renewed importance and increase its reach: the fourfold creates dimensionality and room for thinking by forming an associative poetic opening in sedimented meanings. The gathering of the fourfold, as Heidegger sees in and through the thing, essentially creates nearness and renders visible the four different dimensions of understanding: "the earth" is the basic, dark, and retracted element (Heidegger 2001, 178); "the heavens" is the prominent and lucid element (Heidegger 2001, 178); "the divine" is the bestowing element that by virtue of abundance creates movement and thus exceeds the framework of the apparent (Heidegger 2001, 178); and "the mortals" refer to the humans who, according to Heidegger, are the only beings capable of dying, that is, who are able to understand and relate to an ultimate temporal finitude, as we have seen in the previous chapter (Heidegger 2001, 178).[22] Heidegger names humans "the mortals" because he believes death paradoxically designates the crucial dimension of being human. In other words, *opaque, colored, transgressing, and limiting* epitomize the four dimensions of what things are able to reveal to us; these dimensions capture and determine Heidegger's understanding of the thing.[23] We may also express this insight in a different way: Heidegger understands a thing as a gathering. The exemplary gathering, namely a thing, must in this sense be capable of joining and mediating the seemingly utmost separated dimensions, and it is in this sense that Heidegger's four dimensions are associated with the thing. According to this interpretation of Heidegger's notion of the things, they become the archetypical mediators.

Latour recognizes a number of the prominent dimensions of the concept of the thing and matters-of-concern. As mentioned, Latour holds that it is essential to correct the more or less traditional thing concept in philosophy:

> For too long, objects have been wrongly portrayed as matters-of-fact. This is unfair to them, unfair to science, unfair to objectivity, unfair to experience. They are much more interesting, variegated, uncertain, complicated, far reaching, heterogeneous, risky, historical, local, material and networky than the pathetic version offered for too long by philosophers. (Latour 2005a, 19f)

Latour's distinct concept of the thing expresses several important attributes of things that can be gathered with more comprehensive concepts. And it is exactly in this respect that Latour's concept of the thing appears to be compatible with Heidegger's idea of the fourfold. Concepts in this light are analogous to tools that we can use to understand and gather manifolds in different ways, turning them into specific constellations. The above comparison of Heidegger and Latour also contributes to the demystification of the fourfold and gives an idea of how this concept can be brought into play in relation to our understanding of modern artifacts.

Rereading Heidegger's characterization of the essential dimensions of the fourfold alongside Latour's concept of the thing renders a meaningful and thoughtful connection between the two. The divine in Heidegger's sense is compatible with Latour's concepts of what is uncertain, far-reaching, and network-like; that is to say, Heidegger unfolds the divine as the excessive and transgressing element that is impossible to pin down but insurmountable and always more far-reaching than anticipated. The mortals in Heidegger's terminology are an amalgam of Latour's concepts of risky, interesting, and historical, since human existence is fundamentally exposed and historically defined in Heidegger's view. Precisely due to this limitation, some*thing* may turn out to be remarkable, relevant, or troubling. The earth can be described as local, material, and complex, as it resists being fully understood in general terms—its materiality is solid and compact; the earth's complexity thus stands out because the ability to penetrate it correspondingly decreases. Finally, we may interpret the heavens as the "variegated" and "heterogeneous" in Latour's terminology, because it characterizes what is distinct, manifest, and colorful.

This analogy and its conceptual translation from one context to another are not compulsory; it is, however, associative, possible, and inviting.[24] The analogy shows the room for interpretation and thinking that Heidegger opens up with the idea of the fourfold, which is accommodating and may include Latour's key concepts and *vice versa*. With the concept of the fourfold, Heidegger does not seek to reduce the complexity of the world, on the contrary: by unfolding this concept, Heidegger strives to overcome any fixed or one-dimensional understanding of things.

The analogy does not exclude other translations of Heidegger's concepts, and this translation does not follow with necessity in the sense that it is immediately clear and the only one possible. In this way, the above interpretation should be regarded as an invitation to view Heidegger's concept of the fourfold as a comprehensive, poetic, hermeneutical, and ontological figure for thinking.

Latour recognizes Heidegger's concept of the fourfold but criticizes it for reducing the manifold of things to four (Latour 2004, 235f). As shown in the above interpretation, I consider Heidegger's interpretation of the fourfold to be a fundamental opening, not a limiting reduction of the meaning of things to four. Especially in relation to the dimension of the divine, Heidegger explicitly points to the significant broadening and opening dimension of the fourfold. The concept of the fourfold is indeed a comprehensive concept, like the concept of the thing, but when one unfolds what Heidegger associates with this in relation to *the thing*, then the significance of things is not reduced; on the contrary: things obtain a divine, heavenly glow that makes them ambiguous, fascinating, and complex.

In this sense, the fourfold discloses certain basic dimensions of things and associates the things in the world with humans in an encompassing historical collective—and herein lies one of the many strengths of this concept. According to my interpretation, Latour's critique of the fourfold masks the basic agreement between them. Heidegger would, however, probably agree that he considers some dimensions of things more basic than others, and that he is using the fourfold to try to reveal the most fundamental and ambivalent dimensions in the world that create the best possible space for thinking, thus opening a generative clearing in the middle of the one-dimensional thinking so often connected to technologies. It is this clearing in Heidegger's fourfold that Latour overlooks, but which grants Latour an outlook and perspective on things.

The highlighted link between Heidegger and Latour opens a number of *possibilities* for how the two thinkers can be brought into further dialogue; it also shows how the fourfold may provide a more conscious and multidimensional experience of modern artifacts. The interpretation of Heidegger's concept of the fourfold as a poetic hermeneutic framework separates the present inquiry from Dreyfus and Spinosa's interpretation (Dreyfus and Spinosa 1997, 170f), which attributes, among other things, poetic thinking a less important role in the assembly of the fourfold than the one I advance.

Now it makes sense to take a closer look at the concrete implications of this general hermeneutical interpretation of the fourfold. As mentioned in the first section of this chapter, Heidegger views the airplane as the epitome of modern technology because it prevents nearness and obstructs a gathering of the fourfold:

> All distances in time and space are shrinking. Man now reaches overnight, by plane, places which formerly took weeks and months of travel [. . .] Man puts the longest distances behind him in the shortest time. He puts the greatest distances behind himself and thus puts everything before himself at the shortest range. Yet the frantic abolition of all distances brings no nearness. (Heidegger 2001, 165)

If we take a closer look at the airplane, we find that this artifact also invites interpretations other than the one Heidegger himself forward in the beginning of "The Thing." The airplane makes it possible for humans to suddenly see the earth from a new perspective, and it gathers humans together in its own particular way.

The airplane view, which was surprisingly made available through modern technology in the early twentieth century, fascinated not least the futurist artists, who paid tribute to the airplane as the inaugurator of a new and fascinating era (Marinetti 2013). We can try to imagine how it felt to sit in one of the first modern airplanes as it accelerated and slowly took off from the solid ground below: the unprecedented rush in the stomach and the dramatically changing vibrations in the plane that occurred as it lost contact with the ground and glided through the air. Despite the airplane's exceptionally high speed, at the moment of takeoff it must have felt as if time were standing still as the earth's security and firm grip vanished. Experienced from the airplane, the earth stands out as something that is being left behind. It becomes a massive and solid ground that we took for granted in our everyday lives until we were up in the air. In other words, the massive presence of the earth is disclosed as it withdraws into obscurity. Slowly, a more or less uncertain and unknown space opens up to the pilot and passengers in which bright sunlight and spectacular views are bestowed. As the clouds clear, nothing is standing in the way, and the possibilities and freedom of this unearthly (divine), heavenly dimension are revealed. As humans, we are also vulnerable to countless possibilities in this realm that is suspended between heaven and earth—above all, death. We are vulnerable insofar as we are taking a life-or-death risk just being in this unfamiliar region, but also in the sense that we have been taken out of the framework of our everyday lives on earth and fastened to a specific seat and thus linked to the new technology. The experience of "departing from" belongs to the basic meaning of the concept of existence. "Existence" comes from the verb exist, which means "to stand out" or to "step out of" (Duden 1997, 168). To characterize this fundamental trait of human life that we call existence, a self-reflective relation is required by which this *stepping out of* the familiar and predictable occurs consciously, which is implicit in the above interpretation of the airplane as a gathering of the fourfold.[25]

Another example of how the modern technology of the airplane can move beyond the narrow, instrumental framework that Heidegger warns us about can be found in the following poem by the young American pilot John Gillespie Magee Jr. (1922–1941). In the wake of the futurist movement, Magee wrote a poetic homage to the airplane, which also allows the basic dimensions of the fourfold to stand out:

High Flight

Oh! I have slipped the surly bonds of Earth
And danced the skies on laughter-silvered wings;
Sunward I've climbed, and joined the tumbling mirth
of sun-split clouds,—and done a hundred things
You have not dreamed of—wheeled and soared and swung
High in the sunlit silence. Hov'ring there,
I've chased the shouting wind along, and flung
My eager craft through footless halls of air . . . .

Up, up the long, delirious, burning blue
I've topped the wind-swept heights with easy grace
Where never lark nor ever eagle flew—
And, while with silent lifting mind I've trod
The high untrespassed sanctity of space,
Put out my hand, and touched the face of God. (Magee 2013)

The very title, "High Flight," lets the ambivalence of the ancient myth of Ikaros appear, which is associated with flying. In the excitement of the moment, feeling the magical force of the airplane, the pilot senses a close contact with a divine element. The airplane is, in this interpretation, not just a simple means of transport but becomes a unique vehicle of divine passage. The transport function of the airplane recedes into the background of the poem and gives way to a multidimensional experience of flying. Experienced in this way, the airplane may gather us and shape a space in which we, the mortals, can talk about the fears and fascinations we feel when moving between heaven and earth. We can talk about how it feels to be exposed in this way and about the mediated interconnectivity of the world through a complex network of corridors in the air.

In a traditional reading of Heidegger, it could be argued that not every airplane passenger has a poetic or reflective experience of flying, as Magee did. One possible response to this objection is that not every wine drinker or car driver experiences the jug or the highway as Heidegger does. In a second attempt to respond more fully to this objection, it is not only Heidegger's but also Latour's contention that it requires special attention and training to uncover the significant dimensions of things and to see how a thing is part of a widespread complex network. It requires training to understand and appreciate in which sense the "gods" are also related to the modern airplane.[26] If we understood the fourfold and the essence of things immediately, without the need for poetic mediation or associative thinking of any kind, then the writings of Heidegger and Latour would be redundant. Learning how to identify and understand an artifact in the intriguing light of the fourfold is not a straightforward task. Likewise, understanding and tracking the many associations of things, in accordance with Latour, calls for a special book with an accompanying exhibition of artifacts.[27] The point of departure for such interpretations of specific artifacts may be found in several places. One way to work with this kind of poetic hermeneutics, which is unfolded in relation to the airplane, is to experience firsthand the uncertainties and risks associated with flying, as well as the powerlessness of sitting down and being strapped in an airplane seat. This might generate, according to the above interpretation, links to the divine element. From here, it is possible to see how an airplane is simultaneously a meeting place and an expression of the basic meaning of human existence (the mortals). The habits, traditions, and materials that are often not explicitly thematized in direct relation to the airplane suddenly play a crucial role (the earthly element). This can be seen in conjunction with or in contrast to how airplanes, explicitly and colorfully (the heavenly element), are represented as a rapid and exclusive means of transportation in marketing videos in our Western culture.

When reading Heidegger's and Latour's insights about things and modern technologies, it is important to note another clue that they give us more or less explicitly. Regarding the four dimensions of the fourfold, Heidegger writes, "These four, at one because of what they themselves are, belong together" (Heidegger 2001, 173), and adds, "Each of the four mirrors in its own way the presence of the others" (Heidegger 2001, 179). That is, the four seemingly utmost different dimensions are as such already linked by virtue of their respective meanings. If the meaning of just one of the four dimensions is understood, the meanings of the other three dimensions begin to dawn. It is in this sense that the fourfold may also be regarded as a poetic hermeneutic framework, which in principle can help us understand and relate to *all* things; at the same time, this framework helps create nearness and a gathering around all sorts of things. Likewise, it is the case for Latour's understanding of things that every-*thing* also reflects an extensive network with other things—like *knots* holding the network together. Woven into this network, all things become "interesting, variegated, uncertain, complicated, far-reaching, heterogeneous, risky, historical, local, material and networky" (Latour 2005a, 19f). In the Latourian network, presence and absence are constantly renegotiated.

Finally, it is important to recall a petition and a reservation when interpreting artifacts in the style of Heidegger and Latour. The importance of their pursuits should not be forgotten. Although the claim has been made that Heidegger's thought concerning the fourfold can be expanded and developed to include the epitome of a modern artifact, namely an airplane, it does not mean that all new artifacts and things do not challenge us and threaten to terminate nearness in Heidegger's sense. If we do not try to acquire a fundamental, multidimensional understanding of the many new technological innovations that we come into contact with and interact with in our daily lives, we run the risk of them becoming unimportant and irrelevant to us, and we thus fail to gather around them in any meaningful way and to understand how they may influence our lives.

Viewed from a different but related perspective, many innovations today are produced to meet short-term, one-dimensional needs. Some innovations are practically already outdated as soon as they are put to use. These innovations are, according to Heidegger, the most dangerous, even though they may seem quite harmless. They are mere resources and give rise to the expansion of the standing-reserve. A *thing* is very different from a one-dimensional gadget and cannot simply be produced, since the essence of things is closely linked to our relation to them, as we have seen above. The more or less imminent danger is that these innovations threaten to make our relations to things in the world indifferent and careless. If the human experience of things is indifferent, we become *homeless* in a profound sense (Heidegger 2001). Without understanding things, we no longer have a place to gather, find peace, or focus. As a consequence, we will instead be forced to wander restlessly on earth—more or less consciously—in constant pursuit of the meaning of things. Without things as gathering places, we have nothing to guide us, and beings become mere resources that have lost their integrity and substantial meaning.

According to Latour, things also serve another fundamental purpose. As pointed out above, Latour paints a bleak, almost apocalyptic picture of the contemporary callous relation to things. He believes that if we do not realize the meaning of things and hear their call for attention, we will be handed over to a catastrophic change, which may lead to the end of the world. If we are unable to gather around the things and discuss concrete communal issues in relation to them, then there is no rescue in sight. In other words, in Latour's view, to rescue the earth and our shared destiny with it, we must begin by taking *things* seriously.

The caution expressed in Heidegger's sometimes very abstract and poetic descriptions of the fourfold and in Latour's elaborate understanding of politics and networks is that we must not overlook the thing itself. We should remain conscious of the fact that the fourfold is manifest in relation to the things. Stated differently, the fourfold unfolds differently in different things: the way it is revealed in the airplane differs from how it is manifested in the jug—and each airplane and each jug also give the fourfold a particular configuration. However, it is still the dimensions of the fourfold that are being revealed. There is no question that an abstract understanding merely covers over the things. On the contrary, Heidegger believes that we must do the thinking which unfolds how a concrete (empirical) thing gathers the fourfold in its own special way. If we just use the concept of the fourfold without appreciating the specific things themselves, then there will be no nearness and the things will disappear from our view, just as nearness also threatens to disappear in the incessant production of new things. For Heidegger, it is, however, also the thing that encourages thinking. In Latour's network a similar error is lurking. Are we moving slowly in Latour's network or quickly skipping from one thing to the next? If the latter is the case, the things will dissolve or be mistaken for each other, and gathering and nearness may also risk omission (Latour 2005a, 17f).

Although it is difficult to understand the nature of things, their relations, and different manifestations, we still have to try. For Heidegger as well as for Latour, our existence on earth depends on this endeavor. The uncovering of the specific way of the fourfold in both archaic and modern things unmasks their saving power. Without an understanding of things, humans become *homeless*, according to Heidegger, and Latour maintains that this deficiency can lead to a global catastrophe that destroys our modern habitats. Conversely, if we connect our ideas of things and artifacts, as Heidegger and Latour have prepared us to do, and understand them as basic conditions for our lives, then we can achieve the kind of insight that makes us capable of creating constructive and rewarding communities with all things.

NOTES

1. In this book, I make no systematic differentiation between artifacts and technologies but use the two concepts synonymously.
2. Cf. Riis 2015.
3. In other words, and with reference to Heidegger's later lecture "Building Dwelling Thinking" (2001b): When the dimensions dissolve, there will no longer be a measure, and we will be lost in a fundamental sense and without the means of navigation.
4. It is important to note that Heidegger employs a different thing-concept than he did in *Being and Time*, but it shares some characteristics with his earlier concept of a thing (cf. Heidegger 1996, §15).
5. Heidegger also employs the concept of nearness in *Being and Time* (Heidegger 1996, 98f). Here it is connected to the description of Dasein and in-sein. In *Being and Time*, it is closely connected to the concept of ready-to-handedness, but in contrast to the lecture "The Thing," nearness is primarily overlooked in *Being and Time*, and thus not viewed as a gathering of the fourfold. Heidegger writes in *Being and Time*

> Since Da-sein is essentially spatial in the manner of de-distancing, its associations always take place in a "surrounding world" which is remote from it in a certain leeway. Thus we initially always overlook and fail to hear what is measurably "nearest" to us. (Heidegger 1996, 99)

6. For further explanation of this concept, see Heidegger (1996, 78).
7. Cf. "The nature of the ready-to-hand does anticipate the notion of standing reserve" (Ihde 1979, 124). See also Graham Harman's description of the transition from ready-to-hand to present-at-hand: "In the first instance [ready-to-hand], every object is obliterated, withdrawing into its tool-being in the contexture of the world. In this way, the individual objects are smothered and *enslaved*, emerging into the sun only in the moment of their breakdown" (Harman 2002, 45; emphasis added).
8. This understanding is inspired by Graham Harman but diverges from it insofar as it does even more to highlight the significance of the poetic dimension of the fourfold. Harman writes about Heidegger's use of the concept of the fourfold, "Few clues are given in his writings for interpreting this poetic terminology in a more rigorous theoretical framework" (Harman 2011, 82). As opposed to the more rigorous theoretical framework, I suggest a kind of speculative association that connects the four different dimensions. The difference between the interpretations becomes increasingly clear when Harman writes, "The four poles of the fourfold endorsed by the present book have less poetic names than Heidegger's own. Instead of earth, gods, mortals, and sky, we offer [the following four] real objects, real qualities, sensual objects and sensual qualities" (Harman 2011, 97f). For one thing, and to remain in Harman's terminology, I suggest emphasizing the imagination and highlighting the "imaginative qualities" of the fourfold.
9. "Whatever this fourfold may be, it pertains to the world itself and not merely to the human encounter with the world. It has nothing to do with Kant's Copernican Revolution; it even has the smell of counterrevolution about it" (Harman 2005, 270). See also Latour (1993, 78f).

10. Latour's focus on etymology (part of his methodological toolbox) is also inspired by Heidegger. Furthermore, it is important to notice that in *Making Things Public*, a part of "The Thing" is translated and published, which manifests Heidegger's influence on Latour.

11. See also:

Contrary to what makes Heideggerians weep, there is an extraordinary continuity, which historians and philosophers of technology have increasingly made legible, between nuclear plants, missile-guidance systems, computer-chip design, or subway automation and the ancient mixture of society, symbols, and matter that ethnographers and archaeologists have studied for generations in the culture of New Guinea, Old England or sixteenth-century Burgundy. (Latour 1999, 195)

See also:

But immediately the philosopher [Heidegger] loses this well-intentioned simplicity. Why? Ironically, he himself indicates the reason for this, in an apologue on Heraclitus who used to take shelter in a baker's oven. "Einai gar kai entautha theous"—"here, too, the goods are present," said Heraclitus to visitors who were astonished to see him warming his poor carcass like an ordinary mortal [. . .] "Auch hier nämlich wesen Götter an." But Heidegger is taken in as much as those naïve visitors, since he and his epigones do not expect to find Being except along the Black Forest Holzwege. Being cannot reside in ordinary beings. Everywhere, there is desert. The gods cannot reside in technology that pure *Enframing* [. . .] of being [Ge-stell], that ineluctable fate [Geschick], that supreme danger [Gefahr]. They are not to be sought in science, either, since science has no essence but that of technology [. . .]. They are absent from politics, sociology, psychology, anthropology, history—which is the history of Being, and counts its epochs in millennia. The gods cannot reside in economics—that pure calculation forever mired in beings and worry. They are not to be found in philosophy, either, or in ontology, both of which lost sight of their destiny 2,500 years ago. Thus Heidegger treats the modern world as the visitors treat Heraclitus: with contempt. (Latour 1993, 65f)

12. In this context, confer also: "The object, the Gegenstand, may remain outside of all assemblies but not the Ding" (Latour 2005, 24).

13. Already in *We Have Never Been Modern*, Latour writes about the *The Parliament of Things* and points out the problems that come from not taking things serious enough (Latour 1993, 142f).

14. This theme will be explored further in the next chapter.

15. Cf. Bennett (2010).

16. Latour does not mention this technology, but the example is compatible with Latour's thinking on technology (Latour 2002).

17. See Heidegger's distinction between the right and the true (Heidegger 1977, 313; Riis 2018).

18. To inquire further the fundamental influence of the car on Western civilization, see Kingsley and Urry (2010).

19. Consult also the German original at this point. It is not fully clear how Heidegger envisions the modern autobahn bridge in relation to the fourfold, but based

on Heidegger's own poetic interpretative strategy concerning Hölderlin, Dreyfus and Spinosa's approach is fully legitimate.

20. Cf. Yuk Hui's ambitious search for a multidimensional understanding of modern technology (Hui 2016) and Andrew Feenberg's book *Alternative Modernity* (1995).

21. Cf. Heidegger's notion of "the Other Beginning" (Heidegger 1999b, 125).

22. For a comparable definition of the fourfold, cf. Heidegger (1977b, 351f).

23. By employing a limiting element, such as death, Heidegger gives way to the meaning of "nothingness" in relation to things. This way, Heidegger shows the significance of absence in connection to presence. Negativity shows an insurmountable openness in everything.

24. Cf. Riis (2017).

25. Latour writes,

> Heidegger was not a very good anthropologist of science and technology; he had only four folds, while the smallest shuttle, the shortest war, has millions. How many gods, passions, controls, institutions, techniques, diplomacies, wits have to be folded to connect "earth and sky, divinities and mortals"—oh yes, especially mortals. (Latour 2004, 235)

Here we see that Latour explicitly works with Heidegger's concept of the fourfold but that he wishes to add many more dimensions. Implied in this critique is that Heidegger's four dimensions of things are unequivocal, which differs from my interpretation. Furthermore, I argue that Latour does not acknowledge the significance of concepts as ways to gather the manifold in a certain framework. What is *infolded* in a concept may, of course, be *unfolded*, and in this way, the manifold may reoccur.

26. In this sense, Latour and Heidegger would both agree with Heraclitus. With his reference to Heraclitus, Latour reminds us that it is possible—but requires training—to see the *gods* in modern technology:

> And yet—"here too the gods are present": in a hydroelectric plant on the banks of the Rhine, in subatomic particles, in Adidas shoes as well as in the old wooden clogs hollowed out by hand, in agribusiness as well as in timeworn landscapes, in shopkeepers' calculations as well as in Hölderlin's heartrending verse. (Latour 1993, 66)

Andreas Gursky's photographs of the Rhine (Gursky 1999) are a concrete example of how a work of art can disclose the Rhine just as the hydroelectric plant does.

27. This difficulty says something about the danger we are exposed to and the danger which is characteristic of our time, according to Heidegger.

Chapter 5

Meeting Half Way

As indicated in the introduction, the preceding chapters may be construed as leading up to a discussion on methodology. However, readers less familiar with the works of Heidegger and Latour can commence their reading of this book by delving into this chapter. This chapter focuses on the concept of phenomenology and its relevance to both Heidegger and Latour. At its core, the chapter examines two of their seminal works. In order to provide an initial insight into the methodological reflections of both thinkers, I will begin by presenting excerpts from their texts that explicitly engage with the notion of phenomenology.

> What is important to remember is that bifurcation is unfair to *both* sides: to the human and social side as well as to the non-human or "natural" side—a point always missed by phenomenologists. (Latour 2008, 15)[1]

> The expression "phenomenology" signifies primarily a concept of method. It does not characterize the "what" of the objects of philosophical research in terms of their content but the "how" of such research. (Heidegger 1996, 24)

It is not long ago that Latour published a very ambitious philosophical book that may come to be regarded as his *magnum opus*. It carries the title *An Inquiry into Modes of Existence: An Anthropology of the Moderns* (Latour 2013), and it has not received as much attention as Latour's other books, which may be due to its rather complex composition. In the introductory chapter, Latour writes that he has in fact been working on this book over the past twenty-five years, and now we may investigate it and work together with him along the lines laid out in the book (Latour 2013, xix).[2] With this book, Latour strives to do nothing less than to show how we ought to understand the various sorts of beings in the world, or, better yet, the pluriverse we are living in.

Although philosophers in the history of philosophy were in the habit of envisioning grand projects, such equally ambitious and grand intellectual works are largely missing from contemporary philosophy. In fact, the ambition of Latour's book echoes one of the most disputed and celebrated philosophical books of the twentieth century, Heidegger's *Being and Time* (1927).

In this chapter, I shall argue that these two books are not only equally ambitious but also share a common ambition. Both endeavor to present the foundation for a comprehensive ontology and to inquire into ontology in a *phenomenological* manner. Whereas Heidegger explicitly affirms phenomenology, Latour never describes his method as phenomenological and on more occasions even seeks to denounce Heidegger and phenomenology (Latour 2011, 2013, 37).

To establish these crucial similarities between Heidegger and Latour, or between *Being and Time* (*BT*) and *An Inquiry into Modes of Existence: An Anthropology of the Moderns* (*AIME*), is, however, not the primary aim. Based on this investigation of the two books, we shall get a better grasp of the link between ontology and phenomenology and hopefully be inspired to use their approaches for doing research. The developed connection between *AIME* and *BT* brings both books into a new light and prepares the way for better ontological research.

The structure of this chapter is divided into three parts. In the first part, I present how Heidegger envisions his ontological research in *BT* and makes phenomenology decisive to this enterprise. In the second part, I argue that a phenomenological interpretation of *AIME* not only makes sense, but also captures, connects, and emphasizes Latour's fundamental insights. In the final part of the paper, I will recapitulate Heidegger's and Latour's arguments in a way that clearly indicates the common ground for future phenomenological research. The chapter is the first attempt to compare *BT* and *AIME*. It is not meant to be complete or definitive in its explanation of the two complex works, but rather should encourage further research of two analogous masterpieces of philosophical inquiry.

The terminology of *BT* and *AIME* may pose initial difficulties for some readers and seem incompatible at first sight. Hence, for the proposed comparative philosophy, or *theoretical fieldwork*, one has to be sensitive to the words, pay close attention to the basic conceptions, and not be led astray by different ways of naming them. As we go along, I shall try to explain the key concepts in a way that makes the commonalities of the two books increasingly clear.

HEIDEGGER AND THE WELLSPRINGS OF EXPERIENCE

Heidegger is renowned for his statement, "Ontology is possible only as phenomenology" (Heidegger 1996, 31). He opens his ambitious undertaking in *BT* by turning an ancient perplexity of Plato into a contemporary philosophical concern. Heidegger declares that he wants to give center stage to an ontological inquiry. By quoting from Plato's *Sophist* on the first page of *BT*, Heidegger frames his ontological inquiry as follows:

> For manifestly you have long been aware of what you mean when you use the expression *"being"*. We, however, who used to think we understood it, have now become "perplexed".

> Do we in our time have an answer to the question of what we really mean by the word "being"? Not at all. So it is fitting that we should raise anew *the question of the meaning* of *being*. But are we nowadays even perplexed at our inability to understand the expression "being"? Not at all. So first of all we must reawaken an understanding for the meaning of this question. (Heidegger 1996, 1)

In the thought expressed above, Heidegger articulates modesty in the face of the immense task standing before him: *to first reawaken an understanding for the meaning of the question of being.* He thinks that philosophy in the beginning of the twentieth century is incapable of adequately answering the question of the meaning of being, and, what is worse, that it does not even treat this shortcoming as a serious concern. But the inquiry into an understanding of being is not easy; in fact, it presents such difficulties that *BT* may be seen only as a *prolegomenon* to this undertaking—an undertaking which remained unresolved for the rest of Heidegger's life (Grondin 2001; Kisiel 1993).

In order to better understand Heidegger's approach to ontology, I shall now turn to some of his key thoughts in *BT*. Only two of the originally planned six chapters of *BT* were written and published under that name, and Heidegger had difficulties finishing the project that he himself had anticipated (Grondin 2001, 2f). However, because the introduction to *BT* was written with the grander project in mind, it presents the idea of the entire research endeavor in condensed form (Heidegger 1996, 1ff). Thus, to understand the aim and method of Heidegger's research, particular attention must be given to the introduction. It is divided into two parts: the first part, "The Necessity, Structure and Priority of the Question of Being," addresses the importance of the question concerning the meaning of Being, and the second, "The Double Task in Working Out the Questions of Being: The Method of the Investigation and Its Outline," addresses Heidegger's method of inquiry for the ontological undertaking. It is in this latter part that Heidegger unfolds his conception of phenomenology.

Following Heidegger's thinking in the first part of the introduction, we shall see that his concern for "an explicit retrace of the question of Being" (Heidegger 1996, 2) is significant for understanding his project. As Heidegger writes in the opening lines of *BT*, "This question has today been forgotten—although our time considers itself progressive in again affirming 'metaphysics'" (Heidegger 1996, 2). The question of being was forgotten, but not accidentally. Heidegger sees the missing attention to this fundamental question as inherent in the previous ontological tradition:

> At the beginning of this inquiry the prejudices that constantly instill and repeatedly promote the idea that a questioning of being is not needed cannot be discussed in detail. They are rooted in ancient ontology itself [. . .] We therefore wish to discuss these prejudices only to the extent that the necessity of a retrieve of the question of the meaning of being becomes evident. (Heidegger 1996, 2)

There are three prejudices that Heidegger sets aside to show the importance of the question of the meaning of being. (I) Being is the most universal concept and is always a part of what we understand as beings. According to Heidegger, the universality of being is not that of *genus*, and just because it is always contained in the use of the expression *being* does not mean that its meaning is clear, or "that it needs no further discussion. The concept 'being' is rather the most obscure of all" (Heidegger 1996, 2). (II) The concept of being cannot be defined, which follows from the first conception of being as the most universal concept. Because there is not a more universal genus to which it belongs, it also cannot be understood as a specific being in accordance with traditional approaches to definitions understood in terms of *genus* and *difference*. But just because being cannot be derived from a more universal concept does not mean that it poses no problem to traditional ontology; rather, it shows the clear and fundamental limitations of the traditional logic of definition, which is rooted in ancient ontology. "The indefinability of being does not dispense with the question of its meaning but forces it upon us" (Heidegger 1996, 3). III) Being is an entirely self-evident concept. This third prejudice is connected to the two previous traditional conceptions of being. Similarly, what seems to be a self-evident use of being in fact covers over a lack of understanding that is rooted in the ontological tradition. These preconceptions only readdress the difficulty and significance of Heidegger's inquiry.

> Everybody understands, "The sky *is* blue," "I *am* happy," and similar statements. But this average comprehensibility only demonstrates the incomprehensibility. It shows that an enigma lies *a priori* in every relation and being toward beings as beings. The fact that we live already in an understanding of being and that the meaning of being is at the same time shrouded in darkness proves

the fundamental necessity of repeating the question of the meaning of "being." (Heidegger 1996, 3)

In other words, by showing how not only traditional ontology but also common sense are caught up in misunderstandings and obscurity concerning the most central issue, Heidegger restates the importance of reaffirming the question concerning the meaning of being. And by doing so, Heidegger attains the aim of the first part of the introduction to *BT* (Heidegger 1996, 1f).

Before turning to the second part of the introduction, which deals with Heidegger's methodology, it is important to note that Heidegger also sees the inquiry into the question concerning the meaning of being as serving a positive purpose. As Heidegger states with regard to his own previous exposition

> Up to now the necessity of a retrieve of the question was motivated partly by its venerable origin but above all by the lack of a definite answer, even by the lack of any adequate formulation. But one can demand to know what purpose this question should serve. (Heidegger 1996, 7)

This purpose is presented in relation to different domains of beings, such as "history, nature, space, life, human being, language" (Heidegger 1996, 7), which are thematized as objects by the sciences. These scientific domains and their basic concepts and knowledge are all different domains of beings. Investigating how the fundamental concepts disclose and structure a domain of being as a specific area of being must precede the work of the positive sciences.[3] In other words, one of the main purposes of ontology is to give the different sciences a common framework and point of departure, an idea which we shall also find similarly expressed in Latour's recent work. Heidegger presents this framework as a "productive logic":

> Laying the foundations of the sciences in this way is different in principle from "logic" limping behind, investigating here and there the status of a science in terms of its "method." Such laying of foundations is productive logic in the sense that it leaps ahead, so to speak, into the particular realm of being, discloses it for the first time in its constitutive being, and makes the acquired structures available to the positive sciences as lucid directives for inquiry. (Heidegger 1996, 9)[4]

Heidegger's rearticulation of the question of being aims at an *a priori* understanding of the possibility of the sciences and of the prospect of doing ontology. For these systematic reasons, Heidegger also reconfirms the meaning and prominence of the research he is about to embark upon

All ontology, no matter how rich and tightly knot a system of categories it has at its disposal, remains fundamentally blind and perverts its innermost intent if it has not previously clarified the meaning of being sufficiently and grasped this clarification as its fundamental task. (Heidegger 1996, 9)

After showing the necessity and purpose of the inquiry standing before him, Heidegger addresses his method of inquiry, thus commencing the second part of the introduction of his grand research project (Heidegger 1996, 13f). His reflections on the mode of inquiry concerning the question of the meaning of being are built on insights from the first part of the introduction. The second part consists of three main elements. (I) Heidegger shows the particular status of the human being, Dasein, with respect to ontology; (II) he asserts the need for a destruction of the history of ontology, which is connected to the first part of the introduction; and (III) he links these elements to his fundamental concept of ontological inquiry: *phenomenology*.

Concerning the first element, we must ask why Heidegger starts with an analysis of Dasein. Answering this question is important if we are to avoid misunderstanding Heidegger as a subjectivist or as anthropocentric, and it is also important for the subsequent comparison of Heidegger's and Latour's ontological inquiries. What at first sight appears as a banal fact concerning the question of the meaning of being is how this questioning, this fundamental inquiry, is initiated by human beings; that is, this question is a *concern* for a certain kind of being. Since humans do not occupy a privileged status among beings in a general sense for Heidegger, he does not want to take a subjectivistic turn in the questioning of being. He takes this initial relation between the concern of human beings and the question of being as his point of departure.

> This being, which we ourselves in each case are and which includes inquiry among the possibilities of its being, we formulate terminologically as Da-sein. The explicit and lucid formulation of the question of the meaning of being requires a prior suitable explication of a being (Da-sein) with regard to its being. (Heidegger 1996, 6)

In other words, Heidegger describes the human being as Dasein and characterizes it as being concerned with being and as having the capacity to make an inquiry into being (Heidegger 1996, 10). Heidegger emphatically states, "*Understanding of being is itself a determination of being of Da-sein*" (Heidegger 1996, 10). In its being, Dasein is directed toward its own being and the surrounding world, which gives rise to Heidegger's renowned formula for the basic structure of Dasein: being-in-the-world (Heidegger 1996, 49f). In this sense, Heidegger's fundamental ontological inquiry begins with an elaboration of a particular being in its being, namely Dasein. Dasein has to be accounted for because it interrogates its own being (Heidegger 1996, 6f).

We shall return to these initial and fundamental methodological reflections as we try to understand Heidegger's notion of phenomenology and how it is linked to Latour's thinking in the second and third parts of this chapter. Here we just need to add two further considerations of Heidegger: "The priority of Da-sein over and above all other beings which emerges here without being ontologically clarified obviously has nothing in common with a vapid subjectivizing of the totality of beings" (Heidegger 1996, 12). And it is exactly the preoccupation of Dasein with its own being that foregrounds and connects self-understanding with an understanding of being. "True, Dasein is ontically not only what is near or even nearest—we ourselves are it, each of us. Nevertheless, or precisely for this reason, it is ontologically what is farthest removed" (Heidegger, 1996, 13). This paradox shows itself in the number of available ontologies and anthropologies at hand and must be taken into account as Heidegger sets out on his investigation.

> Not only does an understanding of being belong to Dasein, but this understanding also develops and decays according to the actual manner of being of Dasein at any given time; for this reason it has a wealth of interpretations at its disposal. (Heidegger, 1996, 14)

Not least due to the wealth of interpretations of proper being and the ontology of humans, Heidegger calls for what he describes as a destruction of the history of ontology (Heidegger 1996, 17). The manifold interpretations of being, the many *houses of realism* as Latour would frame it,[5] available to Dasein are, however, only possible because of the constitutional historicity of Dasein "In its factical being Da-sein always is as 'what' it already was. Whether explicitly or not, it *is* its past" (Heidegger 1996, 17). Heidegger thus draws an even stronger connection between the ontology of Dasein and the destruction of the history of ontology. By doing so, he also values the manifold ontologies and shows how they reflect different traditions of experiencing.

Concerning the second element, the manifold traditions of experience, Heidegger maintains that Dasein is so entangled in the world and its own tradition that it neither recognizes its own temporal modifications nor the emergence of dominant ontologies. For this reason, Dasein needs to undertake what Heidegger calls a destruction of the history of ontology.

> What has been handed down it [the tradition] hands over to obviousness; it bars access to those original *wellsprings* out of which the traditional categories and concepts were in part genuinely drawn. The tradition even makes us forget such a provenance altogether. (Heidegger 1996, 19)

These "wellsprings" are the different experiences shaping the tradition. The lack of real genealogical understanding in the tradition makes it unconscious of its own temporal construction process. Just as remote cultures are studied and discussed by anthropologists and ethnographers, if Dasein is to retrieve the meaning of the question of being, it has work to do in its own Western ontological tradition, or more importantly, concerning its own being and basic experiences.

> The tradition uproots the historicity of Da-sein to such a degree that it only takes an interest in the manifold forms of possible types, directions, and standpoints of philosophizing in the most remote and strangest cultures, and with this interest tries to veil its own groundlessness. (Heidegger 1996, 19)

Here, Heidegger encourages what we may read as a call for philosophical anthropology to *come home* and study the fabric of modern society, which at the same time can be viewed as the implicit headline for Latour's *oeuvre*.

The destruction of the history of ontology involves laying bare the genealogy of the traditional and dominant ontologies in order to regain access to the "original wellsprings," that is, to the "original experiences in which the first and subsequently guiding determination of being were gained" (Heidegger 1996, 20). Heidegger's *destruction* is in this sense not negative but affirmative, as he seeks to "stake out the positive possibilities of the tradition, and that always means to fix its *boundaries*" (Heidegger 1996, 20). In doing so, Heidegger highlights how different human practices have in fact influenced the tradition of ontological constructions, thus reawakening the ancient quest of ontology and showing why it is important to pursue its meaning today, that is, why it is always important for Dasein to reassess its own history. In this emphasis on the practical production of ontology, there is another connection to what has become Latour's research program. In the final section of this chapter, we shall return to this shared interest.

Since Dasein is always already its own history, its own tradition shapes the response to the question of being. For this reason, Heidegger claims that ontology can only be carried out as phenomenology, as a reassessment of its own experience, and it is this concept of phenomenology that is the focus of the third and last part of the introduction to *BT* and which sums up the insights expressed above.

Regarding the third element, Heidegger's first approach to phenomenology seems trivial: he divides the word into its ancient Greek components, *phainomenon* and *logos* (Heidegger 1996, 23f). But his interpretation is far from commonplace and gives his mode of phenomenology a specific hermeneutical character, which we shall soon return to. In Heidegger's interpretation, it is important to see that a phenomenon shows itself in human experience; in this sense, it is intimately linked to the concept of experience. However, something may also show itself differently from what it is, "such self-showing we call semblance [*Schein*]" (Heidegger 1996, 25). Phenomenon qua self-showing is, however, structurally connected to phenomenon qua semblance: only because something may also show itself indirectly is semblance in the first place possible, which causes confusion concerning what is apparently alike. Heidegger thereby connects the seemingly opposed meanings of phenomenon. "The confusing multiplicity of 'phenomena' designated by the terms phenomenon, semblance, appearance, mere appearance, can be unraveled only if the concept of phenomenon is understood from the very beginning as self-showing in itself" (Heidegger 1996, 27).

Logos, in contrast, shapes the other part of phenomenology and is normally translated as speech, reason, judgment, concept, etc. (Heidegger 1996, 27f). Heidegger brings together this manifold as he explains the connections between the different meanings. "*Logos* lets something be seen (*phainesthai*), namely what is being talked about, and indeed for the speaker (who serves as a medium) or for those who speak with each other" (Heidegger 1996, 28). In letting something be seen *as* something, logos has a mediating hybrid character and may be viewed as true or false, as a judgment. The implicit theory of truth defines truth as that which makes *correspondence* possible, and it is thus prior to a correspondence theory of truth, which was developed in chapters 1 and 2 of this book. Letting something be understood in relation to something else, disclosing its structural meaning, is a defining feature of Heidegger's concept of truth.[6]

> And because the function of *logos* lies in letting something be seen straightforwardly, in *letting* beings be *apprehended*, *logos* can mean *reason* [. . .] Finally, because *logos* as *legomenon* can also mean what is addressed, as something that has become visible in its relation to something else, in its "relatedness", *logos* acquires the meaning of *relation* and *relationship*. (Heidegger 1996, 30)

In other words, Heidegger uses the concept of *logos* to point to the mediated and relational character of understanding, which means that humans always understand something in relation to something else—they understand something *as* something. What shows itself as something *particular* in a given context or praxis, a phenomenon, becomes associated with other things in *a larger network* when *meaning* is generated, and in this way, the inside view of experience becomes compatible with the outside view. These two moments, standing out and holding together, accompany all experiences and shape the foundation of phenomenology, according to Heidegger. But tradition and habit will always have difficulties distinguishing the two. Phenomena and logos are both constitutional elements of ontology, which we shall find in a slightly moderated and more elaborate form in Latour in the next section.

Taken together, Heidegger claims that phenomenology means "to let what shows itself be seen from itself, just as it shows itself from itself" (Heidegger 1996, 30). When approaching being, there are many possible misconceptions and fallacies: a judgment can be false, things can be concealed or talked about in a way that goes against its traditional categories. Phenomenology needs to be hermeneutical because Dasein is the one practicing phenomenology: in its being, it is concerned with the being of beings, which is shaped by history (Heidegger 1996, 33). Phenomena are always mediated in our understanding inasmuch as we always relate to and interpret phenomena in order to understand them. Dasein thus only shows itself to itself in a mediated form through its different practices.

> From the investigation itself we shall see that the methodological meaning of phenomenological description is interpretation. The *logos* of the phenomenology of Da-sein has the character of *hermeneuein*, through which the proper meaning and the basic structures of the very being of Da-sein are made known to the understanding of being that belongs to Da-sein itself. Phenomenology of *Dasein* is hermeneutics in the original signification of that word, which designates the work of interpretation. (Heidegger 1996, 33)

Fundamental ontology in the Heideggerian sense calls for the phenomenology of Dasein, and this phenomenology defines the horizon of all further ontologies (Heidegger 1996, 33). There is no step beyond Dasein, and our understanding of beings is always relational. For this reason, phenomenology in the mode of Heidegger is interpretive, which is hermeneutical by nature. There are many interpretations of how beings show themselves, but these are self-interpretations. When we learn to see these interpretations in accordance with the related mode of experiencing, we discover the plurality of the *meaning* of beings and come to understand the fundamental framework of all ontologies. In *BT*, Heidegger never gives a definite answer to the question concerning the meaning of being. While this omission is in line with his turn toward Dasein and the plurality of human experience, as well as his primary goal in *BT*, which is to *reawaken an understanding for the meaning of the question of being*, it is not important for the present investigation, which intends to show the fundamental phenomenological similarities in Heidegger's and Latour's modes of research. Based on the analysis above, we are now able to see more clearly the relations between the two thinkers.

LATOUR'S STRUGGLE WITH BIFURCATION

In a language resembling Heidegger's, Latour states, "Yes, there is more than one dwelling place in the kingdom of realism. And each house is built of its own material" (Latour 2011, 323).

Heidegger begins *BT* with a critique of his own time and its inability to come to terms with ontology and the character of human existence. Latour begins his *magnum opus*, *An Inquiry into Modes of Existence: An Anthropology of the Moderns* (*AIME*), in a similar fashion, as he stresses his bewilderment with the contemporary state of affairs and encourages the need to start over, which we shall study further in the next chapter: "'[W]e' no longer know who we are, nor of course where we are, we who had believed we were modern . . . End of modernization. End of story. Time to start over" (Latour 2013, 10). *AIME* is an audacious work and in a certain sense also has the programmatic and unfinished character of *BT*. It is an attempt to *start over* just like *BT*, and encourages others to do the same. *AIME* consists largely of foundational and methodological reflections, as well as Latour's encouragement to readers to contribute further empirical evidence to the philosophical framework mapped out in the book. Implicit in the extensive philosophical and methodological considerations of *AIME* is the assumption that empirical research is important but, in agreement with Heidegger, not self-explanatory. Empirical research does not manifest proper thinking if its framework and the concept of experiencing are not adequately understood, that is, if the hermeneutical dimension—or in Latour's terms, the "interpretative key" of the mode of inquiry—is not justified and transparent (Latour 2013, 56f). And as the subtitle of *AIME* indicates, *An Anthropology of the Moderns*, Latour embarks on an anthropology of who "we" moderns are and defines this in ontological terms. This is indeed similar to Heidegger, who analyzes Dasein in order to define the starting point of ontology.

Neither Heidegger nor Latour sees this starting point as elevating humans to a privileged position among other beings in the world, and both justify their respective points of departure by explicitly stating that *science* and *ontology* are not contingent projects but a specific human practice that must be understood as such, that is, in relation to this kind of being. There are no transcendental subjects in the world but only historical beings for these two thinkers. It makes no sense to speculate about theory independently of human existence and practice; in fact, it is impossible to do so, as such speculation is always already a human practice. Who we are and what we do and care about must be addressed at one and the same time for Heidegger and Latour.

But as Heidegger points out, this also means that human beings are part of their own history, for they are their own history. It was difficult for Latour to develop such an insight, as he struggled with the *modern* worldview, which he did not want to acknowledge as an equally significant part of human history. But in *AIME*, Latour now sees more clearly the modern era as a contemporary condition for research and as a point of departure for his grand research project, which we shall see further developed in the succeeding chapter.[7] Phenomenology in the mode of Heidegger offers no absolute *objectivity* or *subjectivity*. For Latour, the only way to start over is to also take the inside view of the moderns seriously (the *experience* of the various modern practices) and make a comprehensive and comparative anthropology based on different ways of generating meaning in the lifeworld.

A complete exegesis of *AIME* is beyond the scope of this chapter; instead, its purpose is to connect key elements of Latour's work with *BT*. My claim is that Latour develops a kind of hermeneutical phenomenology in *AIME* which defines his ontological ambition and shares fundamental features with Heidegger's mode of phenomenology.

The first clear sign of Latour's phenomenological inclination is his focus on the concept of experience, which is key to phenomenology, as well as pragmatism.[8] On the first page of the introduction to *AIME*, Latour makes experience the leading concept of his investigation: he relies "on the guiding thread of experience, on empiricism as William James defined it: nothing but experience, yes, but *nothing less* than experience" (Latour 2013, xxv). To qualify Latour's notion of experience and see further associations with phenomenology more clearly, it is important to point out that inherent in Latour's concept of experience is a strong critique of the *bifurcation* of subjects and objects, which he underscores time and again (Latour 2011, 2013, 174 and 273). This emphasis gives his concept of experience a clear phenomenological character. This is to say that Latour, like Heidegger, also stresses the fundamental insight in the concept *phainomenon*: when we experience something, when something shows itself to us, it happens as a result of a unification of subjects and objects; *res cogitans* and *res extensa* are inseparable in our primary experience. There is something that shows itself, but it *must not* be disconnected from the perception, involvement, understanding, and feeling thereof; on the contrary, that is, it shows itself in a specific way depending on the mode of access and history of the practitioner. In more general terms, we see how Latour's critique of the bifurcation of subjects and objects connects to his anthropological concern stated above. In these respects, Latour's and Heidegger's modes of phenomenology are analogous.

Latour explains his opposition to and criticism of bifurcation by referencing the classic mistake of dividing the world into primary and secondary qualities, that is, those that are fully independent of humans and those in relation to humans:

> In the seventeenth century, to designate this real, invisible, thinkable, objective, substantial, and formal Mont Aiguille, grasped by the cartography whose practice had been obliterated, people fell into the habit of speaking of its primary qualities—the ones that most resemble the map. To designate the rest (almost everything, let us recall), they spoke of secondary qualities: these are subjective, experienced, visible, perceptible, in short secondary, because they have the serious defect of being unthinkable, unreal, and not part of the substance, the basis, that is, the form of things.
>
> At this stage of reasoning, Mont Aiguille indeed has a double. As Whitehead would put it, the world has begun *to bifurcate*. (Latour 2013, 115; emphasis added)

From a phenomenological point of view, primary and secondary qualities melt together, or better yet, we need to find a language to unify them, so that who we are and what we talk about are addressed at the same time. This language seeks to unify nature and culture into one realm.[9]

To further stress Latour's critique of bifurcation and his mode of phenomenology, he adds that the classic divide between words and things must also be revisited and their mutual history needs to be reassessed:[10]

> To speak of different modes of existence and claim to be investigating these modes with a certain precision is thus to take a new look at the ancient division of labor between words and things, language and being, a division that depends necessarily on a history of philosophy that we shall have to confront, I am afraid, along with everything else. (Latour 2013, 20f)

Language is not a mere secondary quality of things but an integrated part of human practice. In order to realize this, Latour also suggests a kind of destruction of history. Language affects how things present themselves to us humans and, thus, our experience of them—an idea that is also evident in Heidegger's mode of phenomenology and his unfolding of the meaning of logos. To underline the importance of language for Latour and his connection to Heidegger, it is important to recognize that Latour at one point also writes that "words bear their weight of being" (Latour 2013, 21). The philosophical tradition's division of labor between words and things needs to be confronted, and *AIME* strives to show where this tradition goes wrong. The confrontation with tradition as envisioned by Latour is exactly what Heidegger also sought to do with his notion of a destruction of history, which was explained above. Latour asserts that language should instead be seen as an indicator, or better yet, as an integral part of things. Or stated differently, when we give things a name, they are disclosed as a part of human history and exist in a network of meaning.

Taking words seriously in this sense also entails a critique of the correspondence theory of truth, which is another key feature Latour shares with Heidegger. In this context, it is also interesting to note that Latour, in a text leading up to *AIME*, modified an early phenomenological motto, "[Go] back to the object" (Latour 2004b, 66), which comes quite close to the initial dictum of phenomenology by Edmund Husserl, the teacher of Heidegger, who wrote, "Go back to the things themselves" (Husserl 2001, 168).

Combining phenomenological insights with the pragmatist tradition, Latour translates absolutist statements about truth (with a capital "T") into a question of "verification" and a description of "felicity" and "infelicity" conditions for passing and assessing judgments, hence embedding the notion of truth into a practical context (Latour 2013, 20f). "It turns out, though, that there are several types of truth and falsity, each dependent on very specific, practical, experimental conditions. Indeed, it can't be helped: there is more than one dwelling place in the Realm of Reason" (Latour 2013, 18). In other words, Latour is interested in identifying and describing the values associated with specific conventions, communities, and institutions as they pass, accept, and work with judgments and performative statements—as they experience the world and process it in praxis. The values associated with the acceptance or rejection of a judgment are what interest Latour in connection to felicity and infelicity. By understanding the community's values, and what they care about, that is, the *felicity* and *infelicity* conditions, Latour softens his critique of the moderns, as he must admit that they are also to be taken seriously, and it would be a misunderstanding to suggest a general critique of them (Melson and Riis, 2015).[11] It is crucial to see human beings in relation to their dwelling and account of ontology, that is, the inseparability of the mode of existence of human beings and their world must be understood, which is expressed with the concept of a lifeworld in the phenomenological tradition.

By emphasizing the role of what humans care about, Latour turns the meaning of truth into a description of the governing principles of a specific praxis (defined by certain values). Such principles are more or less explicitly manifest in the justification of the acceptance or rejection of claims of knowledge in practice. Values are understood here as the interpretive key (Latour 2013, 48), since they define a domain from within. Values are the norms of a community; they determine how the world is seen and what is cared about—they supply the measure and define the center of concern. In this sense, values are the key to understanding the particular interpretations of the world and our experiences in it. This does not mean that our practices are contingent. On the contrary, practices function and are accepted not because of their *objective correspondence* with the world but because they, based on pragmatic and finally normative criteria, shape what we care about. Stated differently, in Latour's view, we may speak of *truth values* in a more general and nuanced sense: they are not just true or false but define what is accepted, effective, okay, overlooked, criticized, rejected, etc., within different communities and practices of truth production. Values are expressed, so to speak, in the preposition, or form the basic attitude, they provide "the key in which what follows is to be interpreted" (Latour 2013, 62), that is, the mode of access.[12] The idea of values as prepositional attitudes toward the world is hence reminiscent of Heidegger's definition of phenomenon as something that presents itself at first and second sight. Latour illustrates the fundamental role of values, which becomes evident when we make mistakes or confuse one thing for another:

> The canonical example involves a foreign visitor going through the buildings of the Sorbonne, one after another; at the end of the day he complains that he "hasn't seen the University of the Sorbonne." His request has been misunderstood: he wanted to see an institution, but he has been shown buildings . . . For he had sought in one entity an entirely different entity from what the first could show him. He should have been introduced to the rector, or the faculty assembly, or the institution's attorney. His interlocutors had misheard the key in which what he was requesting could be judged true or false, satisfactory or unsatisfactory. (Latour 2013, 49)[13]

To understand the nuances of the two meanings of university, we need to understand what the visitor cares about, or his values. This is the interpretive key—the hermeneutical framework—that will allow us to meaningfully interact with him and respond to his query satisfactorily. To insist that he has in fact seen the university when he has only seen the buildings would be to make a "category mistake" in Latour's sense, as two sets of values and two different domains have been confused. For experience to make sense, we need the adequate interpretive key, the precondition, which is prior to the experience itself.

The Sorbonne example shows that there is a difference in attitude concerning what we care about, or between different values. The visitor's request arises from a certain interest, and we must be able to understand that interest in order to help him. For Heidegger, the caring structure is also a defining moment of human beings (Heidegger 1996, 118f),[14] which highlights intentionality as another common phenomenological feature of Latour and Heidegger. Humans are always already preoccupied or concerned with certain things, emphasizing the aforementioned fundamental connectedness to the world. Whether we are inquiring, helping, judging, criticizing, sensing, and so on, we approach the world in a way that expresses our values and what we care about.

But the borders of practices and domains dominated by certain values are not absolute, according to Latour. What and how we care about something bypasses and breaks down borders all the time. In one of Latour's examples, the faculty assembly assembles in one of the university buildings and thus shows how the two meanings above are indeed also connected. To describe this mobility across boundaries, across the different domains we care about, Latour uses the concept of a "network," as pointed out in earlier chapters. In order to better understand this Latourian key term, we may take a look at one of its most apparent and controversial forms: a network may be understood as a gas pipeline system such as the one going from Russia through the Baltic Sea to Europe or through Ukraine to Europe. Based on this notion of a network, we can literally see how a network crosses borders. In less obvious but still very evident ways, gas, Ukraine, risotto, and warfare—seemingly very different entities—may be associated in one and the same network when we cook. This ability to connect seemingly very different things is one of the key strengths of Latour's notion of a network:

> Everyone notices this [that a gas pipeline consists of a number of different actors], moreover, when some geopolitical crisis interrupts gas deliveries [. . .]. Everyone then sets out to explore *all over again* the set of elements that have to be knitted together if there is to be a "resumption of deliveries." Had you anticipated that link between the Ukraine and cooking your risotto? No. But you are discovering it now. If this happens to you, you will perhaps notice with some surprise that for gas to get to your stove it had to *pass through* the moods of the Ukrainian president . . . Behind the concept network, there is always that movement, and that surprise.[15] (Latour 2013, 32f)

At the same time, it is noteworthy that "network" in the example above has two different meanings, as it may describe both the fixed pipeline system and what is circulated, that is, the gas: "[U]nder the word 'network' we must be careful not to confuse what circulates *once everything is in place* with the setup involving the heterogeneous set of elements that allow circulation to occur" (Latour 2013, 32).

For the protagonist of *AIME*, a female anthropologist trying to understand the moderns, the concept of network is very useful, as it relativizes the seemingly absolute boundaries of established domains of knowledge production. Furthermore, this protagonist also gives Latour a chance to play with first-person and third-person perspectives in the book.

> It must be acknowledged that the discovery of the notion of network whose topology is so different from that of distinct domains, gives her great satisfaction, at least at first. Especially because these connections can all be followed by starting with different segments. If she chooses to use a patent as her vehicle, for example, she will go off and visit in turn a laboratory, a lawyer's office, a board of trustees, a bank, a courthouse, and so on. But a different vehicle will lead her to visit other types of practices that are

just as heterogeneous, following a different order on each occasion. (Latour 2013, 30f)

Latour uses the concept of "values" to define and disjoin domains and the notion of "networks" to connect them. Values support discontinuity between domains, whereas networks show continuity across seemingly different spheres. In Latour's previous work on networks in Actor-Network-Theory (ANT), the importance of values was set aside in favor of networks, which overcome boundaries.[16] Understanding networks and values, or the continuity and discontinuity between entities and institutions, in an egalitarian way is one of the key qualities defining *AIME*. This feature also leads to a different and more nuanced understanding of the moderns, which I shall return to in the next chapter.

The goal of Latour's "inquiry into modes of existence" could be defined as an investigation of the relation between networks and values: how networks (abbreviated by Latour with [net]) connect domains with different values and propositions (abbreviated by Latour with [pre]). Latour wants to be able to identify distinct ways of knowledge production and different (regional) ontologies in their relations to each other in such a manner that they do justice to and are compatible with the values and propositions of the practitioners themselves (i.e., their own first-person experience); at the same time, he seeks to fit these various experiences into a comprehensive framework, where each practice becomes a specific position (third-person perspective).[17] Stated differently, Latour strives to connect the outside view of the philosophical anthropologist with the inside view of the practitioners.

> I can now recapitulate the object of this research. By linking the two modes [NET] and [PRE], the inquiry claims to be teaching the art of *speaking well* to one's interlocutors about what they are doing—what they are going through, what they are—and what they care about [. . .]
>
> TO DESCRIBE networks in the [net] mode, at the risk of shocking practitioners who are not at all accustomed, in modernism, to speaking of themselves in this way; TO VERIFY with these same practitioners that everything one is saying about them is indeed exactly what they know about themselves, but only in practice; TO EXPLORE the reasons for the gap between what the description reveals and the account provided by the actors, using the concepts of networks and prepositions.
>
> Finally, and this is the riskiest requirement, TO PROPOSE a different formulation of the link between practice and theory that would make it possible to close the gap between them [theory and practice][. . .]. (Latour 2013, 64f)

Understanding what the practitioners care about—their values, what they are going through, and the network they are part of—defines the *crossing* of the two most fundamental modes of existence in Latour's ontology. Each domain, each praxis, and each mode of existence has its own characteristics, which Latour wishes to explore and explain in accordance with this fundamental crossing. The isolation of domains is, on the other hand, counterbalanced by the exploration of how domains establish contact and cross the borders into other domains, as each domain belongs to various networks. In this way, law, science, and religion are specific domains with their own key concepts, values, and felicity conditions, and each domain accounts in its own way for the beings in the world. At the same time, the domains are connected to each other in multiple ways through networks spanning people, artifacts, schools, cities, and actants of all sorts.

It is in this context that Latour summons the ancient Greek notion of logos, which also played a crucial role in the goal of Heidegger's ambitious research project, "After all, isn't wanting to speak well about something to someone, standing in the agora, a fairly good approximation of what the Greeks called *logos*" (Latour 2013, 66). More specifically, the multiverse of things should be understood as it is *experienced*: this plurality should not be confused, and the different practitioners should not be mistaken for one another (Latour 2013, 262f and 280).

Within the Latourian framework, ontology is multimodal and thought from within the experiences of the practitioners (Latour 2013, 181f). The things the practitioners care about, their values, and core concepts should be accommodated and respected. Latour also strives to find a common language, or an outside meta-view, in which it is possible to understand the regional ontologies as domain dependent and as the result of a specific involvement of humans and nonhumans; the commonalities among these regional ontologies are to become transparent and connected. In this sense, Latour seeks to describe ontology in phenomenological terms as relative to human practices, just as Heidegger does. Unlike Heidegger in *BT*, Latour seeks to specify the interpretive key of each domain and thus to elaborate on the internal differences among practices. However, these sub-ontologies still reflect the same basic phenomenological embeddedness of humans in the world, the priority of praxis, and the relativity of truth production, which Heidegger also stresses in *BT*; they differ because of the many different ways humans are embedded in the world and the associated truth-in-the-world. The following passage is now clearer

> We see why the expression "to each his own (truth)" not only has the relativist tonality people often grant it; it also implies the daunting requirement of knowing how to speak of each mode in its own language and according to its own principle of veridication. (Latour 2013, 143)

If we can learn to understand that this relativity of truth is in light of the various experiences from within a given praxis, and that it has clear limitations depending on what we care about, then we can also see how the fundamental dimensions of Latour's project in *AIME* are analogous to Heidegger's in *BT*.

THE MIDDLE KINGDOM

In the final part of this chapter, we shall rearticulate and highlight the key findings from the interpretations above and thus learn to see more clearly the links between Heidegger's and Latour's modes of phenomenology.

> If we seek to deploy the Middle Kingdom for itself, we are obliged to invert the general form of the explanations. The point of separation—and conjunction—becomes the point of departure. The explanations no longer proceed from pure forms toward phenomena, but from the centre toward the extremes. (Latour 1993, 78)

In the section above, it became increasingly clear that Heidegger and Latour share a number of significant phenomenological insights. This is manifested, for instance, in their endeavors to take seriously the experiences of practitioners—thinking from the center of experience and subsequently about humans and nonhumans. Without overruling them, these experiences must be integrated into a common framework of praxis. By articulating the lifeworld as the "kingdom in the middle" of two extremes—between the bifurcation of subjects and objects and the anthropology that negates their differentiation—the concept of experience is defined and strengthened, and the phenomenological mode of inquiry is reconfirmed.

For Heidegger and Latour, the world shows itself differently for different beings depending on how they engage with it. Mediation and felicity conditions become key elements, as well as a deflation of the concept of truth in accordance with particular unveiling processes and specific truth production practices. But this also engages ontology and how it is supposed to be carried out. In one sense, this approach generates an ontological pluralism (Latour 2013, 21), as there is no way to go beyond the different modes of experience, as each unveils a different mode of being that cannot be separated from a mode of human existence. In another sense, this shared approach to ontology, which emerges from a shared view of the central role of experience, allows Latour and Heidegger to investigate what holds the manifold of human experiences together and to create general guidelines for ontological inquiries.

The conceptual analogy between Heidegger and Latour may be taken one step further. Heidegger's definition of phenomenology can be compared with

Latour's basic *crossing* between *prepositions* and *networks*. Heidegger sees phenomenology as the science of phenomena in the sense that a phenomenon is the mode in which beings present themselves to human beings. But, as Heidegger emphasizes, beings present themselves differently according to how we approach them, that is, according to different practices and the specific concerns of human beings "Beings can show themselves from themselves in various ways, depending on the mode of access to them" (Heidegger 1996, 25). It is this insight we must bear in mind: priority should be given to experience in phenomenology so that we can describe the exact circumstances and modes of access generating various experiences, that is, give a detailed account of a certain phenomenon. Conversely, we should also be able to show how key anthropological concepts are rooted in experience, thus countering "all free-floating constructions and accidental findings" (Heidegger 1996, 24). In generating this exact description, the logos or science of the phenomenon is needed. However, in describing something *as* something else, a translation or an interpretation takes place, which binds the being showing itself to a larger network of meaning and to other beings as well. This is why Heidegger declares that the logos of phenomenology is hermeneutical, that is, interpretive (Heidegger 1996, 33). This logos is at first sight domain dependent, but when inquired into thoroughly, it shows familiarity across domains, just as one of the basic meanings of logos is relatedness. In the same vein, Latour also thinks that each praxis makes beings appear in a particular way according to the concerns of the praxis, or its defining values and prepositions. When seeking to understand the experience of certain praxes or particular communities and how beings present themselves in these contexts, Latour also underlines the necessity of a detailed understanding of its mode of access, which ultimately depends on its defining values.[18] However, in order to connect the different experiences and practices, Latour also calls for practical hermeneutical work to show the various common networks of the practices. Seen from this perspective, the main difference between Heidegger and Latour is their primary concern. Heidegger's primary concern is the meaning of being as it is conveyed through different human experiences, whereas Latour is preoccupied with understanding the framework of different human experiences of beings. In spite of this difference, Heidegger and Latour both value the relativity of the human experience of beings and ascribe this relativity to the diversity in the mode of being that we ourselves are.

Hence, Heidegger and Latour point to a common phenomenological anthropology as the best way to do ontology: a shared understanding of how humans are in and engage with the world. There are no *a priori* limits to the number of different experiences humans may obtain. Human experience, for both philosophers, is always guided by what we humans care about, and this caring structure ties us to our surroundings and links the things we care about together in specific ways. The caring structure highlights particular relations, that is, certain networks are unveiled as a consequence of what we are directed toward in our caring. Heidegger and Latour see the *meaning* of being not as unequivocal but as relational and as what is brought forth in our praxis. As such, this relational, revealing structure shapes the non-absolute foundation of ontology and the meaning of being.[19]

That we humans are not necessarily fixed to one set of values, that we are not tied to only one specific set of relations and are able to engage with the world in numerous ways, gives us an openness that manifests something profound about the sort of beings we are. Humans are able to see and engage with various ways of unveiling beings through science, religion, history, art, etc., and this manifold points to an openness in the relation between humans and their worlds. This openness is furthermore also what we may understand as human freedom. According to Heidegger and Latour, it is the same openness that allows humans to act as "shepherds" of being (Heidegger 1977d; Latour 2011). As shepherds of being, humans should take care of and talk well about beings in a way that supports the manifold and holds the various interpretations of being together.

By unfolding this reciprocal phenomenological interpretation of *BT* and *AIME*, we can now see Heidegger and Latour as two different thinkers with a common goal: to develop a fundamental ontology based on a phenomenological methodology. The two complex books and associated modes of phenomenology need further inquiry, but based on the interpretation above, I hope to have delivered the basic framework within which this research may take place.

NOTES

1. How did we manage to behave as if Nature had "bifurcated" into primary qualities—which, if you remember, are real, material, without values and goals and only known through totally unknown conduits—and secondary qualities which are nothing but "psychic additions" projected by the human mind onto a meaningless world of pure matter and which have no external reality although they carry goals and values. How did we succeed in having the whole of philosophy reduced to a choice between two meaninglessnesses: the real but meaningless matter and the meaningful but unreal symbol? (Latour 2008, 36)

2. Confer also the online collaborative project: http://www.modesofexistence.org/

3. Such an investigation also belongs to what I have described as theoretical fieldwork in the introduction to this book.

4. Outside the scope of this book, but philosophically worthwhile, would be to compare what Heidegger here calls a "productive logic" and what Thomas Kuhn

develops much later with the concept of a paradigm. In this respect, further inspiration may also be attained in chapter 1 of this book.

5. Cf. Latour (2011, 323)

6. Cf. Heidegger's critique of the correspondence theory of truth:

> Furthermore, because *logos* lets something be seen, it can *therefore* be true or false. But everything depends on staying clear of any concept of truth construed in the sense of "correspondence" or "accordance" [*Übereinstimmung*]. This idea is by no means the primary one in the concept of *aletheia*. The "being true" of *logos* as *aletheuein* means: to take beings that are being talked *about* in *legein* as *apophainesthai* out of their concealment; to let them be seen as something unconcealed (*alethes*); to *discover* them. (Heidegger 1996, 29)

7. Heidegger also criticized the moderns but took them very seriously at the beginning, in opposition to Latour (cf. Riis 2018).

8. In this chapter, we shall not explore further the obvious connections between Latour and the pragmatist tradition or the relation between phenomenology and pragmatism.

9. In a preliminary work to *AIME*, "Reflections on Etienne Souriau's Les différents modes d'existence," Latour explicitly attacks phenomenology, but in such a way that he, in fact affirms it in the sense developed by Heidegger.

> This phenomenon is the polar opposite of that found in phenomenology. With wicked humour, Souriau cites Kipling: "in the end phenomenology is where one is least likely to find the phenomenon . . . as in Whitehead, Souriau's phenomenon is no longer caught in a pincer movement between what might be behind it (primary qualities) and what might be ahead of it (secondary qualities)." (Latour 2011, 18)

Doing phenomenology, according to Heidegger, is to criticize notions such as primary and secondary qualities; after all, Dasein and its existence in-the-world are all there is.

10. Confer also Heidegger's concept of destruction (Heidegger 1996, 17f).

11. However, Latour's confrontation with modernity and his ambivalence toward it also resemble Heidegger's late philosophy.

12. Cf. Latour (2013, 57).

13. Heidegger also works with the same model of understanding through the "break-down" of expectations and praxis; compare his analysis in *Being and Time* (Heidegger 1996, §16).

14. Cf. Verbeek (2008).

15. See also Heidegger's broken tool analysis (Heidegger 1996, §16). Both Latour and Heidegger explain how breakdowns make visible a network of equipment, or the *Zeugganzheit*. Cf. Galison (2006).

16. Latour writes,

> This theory [ANT] played a critical role in dissolving overly narrow notions of institutions, in making it possible to follow the liaisons between humans and nonhumans, and especially in transforming the notion of "the social" and society into a general principle of free association, rather than being an ingredient distinct from the others. Thanks to this theory, society is no longer made of a particular material, the social—as opposed,

for example to the organic, the material, the economic, or the psychological; rather, it consists in a movement of connections that are ever more extensive and surprising in each case. And yet, we understand this now, this method has retained some of the limitations of critical thought: the vocabulary it offers is liberating, but too limited to distinguish the values to which the informants cling so doggedly. It is thus not entirely without justification that this theory is accused of being Machiavellian: everything can be associated with everything, without any way to know how to define what may succeed and what may fail [. . .] In this new inquiry, the principle of free association no longer offers the same meta language for all situations; it has to become just one of the forms through which we can grasp any course of action whatsoever. (Latour 2013, 64)

17. This grand encompassing capacity of Latour has also led contemporaries to associate him with Hegel: "One of the great intellectual adventures [Latour] of our epoch [. . .] the Hegel of our times," enthuses Maniglier (Maniglier 2013).

18. Also underscored by his numerous field studies; see especially Latour (1979, 1992).

19. This version of phenomenology has also inspired the school of postphenomenological thinking, which in particular focuses on the technical mediation of experience (Ihde 1995, 2010; Verbeek and Rosenberger 2015).

Chapter 6

The End of the World as We Know It

Over the past years and up until his recent death, Bruno Latour had been confronted with a number of new challenges, difficult questions, and thought-provoking paradoxes. "So, the question of what it means for a people to live in space, on land, from a soil is wide open all over again" (Latour and Weibel 2020, 6). If we never have been modern, as he was claiming by the end of the twentieth century—if there has never been a rupture or transformation indicating a *modern epoch*—then why are we now facing global climate change and environmental catastrophes?[1] Why does Latour at the same time contend that "nothing will be as it was before"? (Latour and Weibel 2020; Latour 2018, 17). In this light, Latour's recent claim that "we have reached the end of a certain historical arch" (Latour 2018, 32) is also paradoxical if he does not take seriously the distinct historical period of modernity.[2] How can Latour's investigations and conceptual framework make the end of an era sufficiently clear, highlight the immense importance of its cessation by claiming that *everything changes*, and emphasize continuity in history? Finally, who is the "we" Latour addresses, who stands at the end of a historical arch? Is it the same "we" who created the problems, the one reflected in Latour's claim that "we have to live with the consequences of what we have unleashed" (Latour 2018, 20)? Or is it the "we" from "We have never been modern"? For one thing, it seems certain that the late Latour is pointing his finger at humans and thinks that "we" carry a special kind of responsibility, which suggests an asymmetry between humans and nonhumans.

It is becoming increasingly clear that the late Latour began taking the radical environmental problems of our time seriously and that he thought humans had generated an enormous impact on the planet over the past centuries. It may always be difficult to discern the beginning of a development or to assess its accumulative effects in an early stage. At a later point in time, however, the compound effects may be almost impossible to neglect. This was also the case for Latour concerning the dangers of modernity and global climate change. The arch that is coming to an end, which Latour refers to, is only the very late effect of something that happened much earlier, the arch's beginning, but which Latour has been awfully late to recognize and counter.[3] In his final works, he does, however, show an increasing awareness of the modern origin of the present-day dangers to life on Earth. Now, the end of the so-called civilization appears to be close at hand to him. He has come a very long way since he ridiculed modernity and analyzed the hole in the ozone layer from his *armchair* while reading a newspaper.[4]

In *We Have Never Been Modern*, which was originally published in 1989, his primary interest concerning the ozone layer was to show this phenomenon as a contested object and a manifestation of a so-called "hybrid object." He was not interested in revealing an alarming and specific modern threat to life on Earth—even though the latter should have been a warning sign that something highly dangerous was taking place and threatening life on Earth. At the end of the twentieth century, Latour emphasized the continuity of time, the equal responsibility of humans and nonhumans, and that the sort of challenge one is confronted with in defining the hole in the ozone layer in fact resembles previous (pre-modern) problems in all for him relevant aspects. In Latour's terms, the hole in the ozone layer was just an expression of an archaic thing, namely a hybrid object, not an alarming modern mutation of the atmosphere.

Fast forward two to three decades after writing *We Have Never Been Modern*: Latour realizes the enormous harm done to life on Earth by what he, for the most part, has chosen to call "the modernizers" (Latour 2017). This terminology allowed him to avoid the notion of "modernity" and its dangers.

In clear opposition to such an understanding of modernity, Heidegger did not try to circumvent the notion of modernity but insisted on it. Most of his mature thinking is devoted to warning his readers and listeners against its dangers. Latour did not hesitate to ridicule Heidegger's efforts, as we have seen above. But when Latour woke from his lightheaded "comedian slumbers," he acknowledged in his last writings the serious tragedy of modernity.[5]

To appreciate Latour's legacy, it is nonetheless important to point out that he died while fighting what he recognized as the great war of the twenty-first century—one that even dwarfs the impacts of the previous world wars of the twentieth century (Latour and Weibel 2020, 5). This war calls for new arms, the redrawing of previous front lines, and a reevaluation of all human values. To prepare for this fight against the modernizers and for the Earth, Latour finds hope in the arts, just as Heidegger did nearly a century before him (Riis 2018). It is my claim, and the thesis I want to unfold here in the final chapter of the book, that Latour's struggle against modernity and emphasis on art and architecture bring his thinking a final step closer to his philosophical predecessor Heidegger, and that this proximity may come to define his legacy.

CLEAR AND PRESENT DANGER

To describe the challenges that Latour is faced with—that humanity in general and all terrestrial beings are faced with—brought about by the changing Earth, he frequently uses metaphors of warfare, "And yet it's a war for good, a war of extermination, no question about that—and of planetary dimensions" (Latour and Weibel 2020, 5). And he continues, "As if we felt attacked everywhere, in our habits and in our possessions" (Latour 2018, 8).[6] "So, are we faced with civil wars? No, with something much worse, because each combatant is divided inside itself as well" (Latour and Weibel 2020, 5).

To further underline the scale and gravity of what is taking place at this historic moment, Latour also draws an analogy to the Scientific Revolution: Just as it was the case four centuries ago, the role of science, politics, and religion shall once again be fundamentally transformed (Latour and Weibel 2020, 3f).[7] The inhabitants on Earth are once again being attacked and disoriented in three profound yet different ways: *in space, in time, and in identity,* just as they were during the Scientific Revolution (Latour and Weibel 2020, 3). The earthlings do not know where they really are, in which period they actually live, or how to make sense of themselves, that is, to define what agency they really have and what they ought to do. The confusion is extreme: the very place that the modern sciences had taken for granted as a stable framework and ground, the Earth, has been unmistakably shaken.

Understanding the Earth as a stable, homogeneous globe is, however, not least a product of modern science, according to Latour. As he points out, "Haven't you ever worried that when you say that the Earth is a planet, that it is a globe, you actually have to mentally position yourself as if you were considering it from out in space?" (Latour and Weible 2022, 3). It is in this sense that the projection of the Earth as a planet floating quietly and timely around the sun is a product of the Scientific Revolution, which was initiated by Galileo and Newton and their twentieth-century followers, who photographed the Earth from space and made the *blue planet* an icon.

> And yet this vision from the vantage point of the universe—"the view from nowhere"—has become the new common sense to which the terms "rational" and even "scientific" find themselves durably attached. From now on, it is from this Great Outside that the old primordial Earth is going to be known, weighed, and judged. (Latour 2018, 68)

This planetary vision of the Earth neither assesses nor reveals any of the radical climate and environmental problems taking place on the thin surface of the Earth; on the contrary (Latour 2018, 67). To view the Earth from the outside reduces our sensitivity to the extermination of various species, weather catastrophes, famines, migrations, and all the radical changes following climate change and environmental disasters. Instead, this vantage point leaves the misleading impression of stability, which influences planetary politics. By framing the habitat of living species on Earth from an outside location, too much distance has been created to awaken its inhabitants from their dangerous slumber. With the advance of climate change, however, the Earth is now intruding into all worldly affairs and is about to change politics, science, and the arts, as well as our presumable private lives for good.

The trinity of Earth, Globe, and Nature has shaped modern metaphysics, which has successfully advanced present-day sciences and technological development, but is ill-equipped to understand the terrestrial transformations of "genesis, birth, growth, life, death, decay, metamorphoses" (Latour 2018, 68). Because of this incapacity and metaphysical disinterest, environmental and climate crises have been able to develop in the background of the seemingly important advancement of modern sciences; they have gone unnoticed until becoming so evident that they now impact and threaten all life-forms on Earth.

In this sense, Latour's project can be described as manifesting the dangers of modern metaphysics, which were generated and enhanced by the Scientific Revolution and associated with the beginning of modernity. Subsequently, the objective is to develop a different and more adequate concept of Nature and our particular planet Earth. While the early Latour first conceived this project as one that would correct epistemological errors and misunderstandings, the late Latour has come to understand that overcoming (the metaphysics of) modernity is urgent and indeed a matter of life and death. That "we" have been modern for far too long was, in a way, one of his last essential insights.

A significant and associated danger of the modern worldview is that globalization has become an ideal in the sense that differences are smoothed out in order to align with worldwide standards. As Latour states, "The battle cry 'Modernize!' has no content but this: all resistance to globalization will be immediately deemed illegitimate" (Latour 2018, 14). To globalize and to standardize in practice means to make the world still more compliant with the view of the Earth from the outside—with a single metric that pushes for uniformity. Latour reminds his readers, "We should never forget that a globe is never bigger than the screen (or piece of paper) on which it is spread" (Latour and Weibel 2020, 3). This is to say that the scale and distance of our maps have concrete consequences for our understanding of the field we want to inquire into and understand, and that the scalable maps have become a problem in terms of how we respond to the new sort of locally reflected problems confronting us.

If we combine this view of planet Earth with a concept of Nature that has been playing a key role since the Scientific Revolution, then it becomes difficult to understand the profound influence of the Earth on politics and *vice versa*. This dualistic worldview *a priori* makes it difficult to understand the hybrid connection between the Earth and politics, which is needed in order to grasp the gravity of the problems all living beings are confronted with and the sort of responsibility that modern humanity is facing. As a consequence, Latour encourages his readers to resist the epistemological framing of the modern sciences

> [I]f we swallow the usual epistemological whole, we shall find ourselves again prisoners of a conception of "nature" that is impossible to politicize since it has been invented precisely to limit human action thanks to an appeal to the laws of objective nature that cannot be questioned. (Latour 2018, 6)

This may also be read as an affirmation of the claims developed in chapter 2. However, it is important to notice the implicit asymmetry between humans and nonhumans necessary for categories like politics or responsibility. For either politics or responsibility to take place, humans must learn to see themselves differently than objects; they have to be aware of the fact that the problems at stake are the failure of politics, as well as the failure to take responsibility.

HOW DO WE GET DOWN TO EARTH?

Given the rapid changes on every scale and dimension of the surface of the Earth, everybody, all terrestrials, will be more or less effected and disoriented. To avoid becoming fully lost, it is important to try to recalibrate the ways in which the Earth can be navigated. The bewildering changes call for imaginative poetic thinking and for the development of new sensitivities. This creates a privileged time for philosophy and art. "In a quieter period, it might make sense for scientists to reject the collaboration with artists, or to limit their help to decoration and popularization. Not in a time of crisis such as that of the newly moving Earth" (Latour and Weibel 2020, 7).

For Latour, the big question of our present epoch becomes how to appropriately respond to the changing Earth and find a place to dwell. How can we get down to and inhabit the Earth instead of drifting homelessly and unattached around the globe? As Latour (self-)critically remarks, "And thus no one has the answer to the question 'how can one find inhabitable land?'" (Latour 2018, 16). He resumes, "We don't know where to go, or how to live, or with whom to cohabit. What must we do to find a place? How are we to orient ourselves?" (Latour 2018, 16). There is no escape from Earth—and a profound *earthquake* is shaking the world in slow motion. What are the terrestrials supposed to do now? (cf. Latour 2018, 17).

For several years now, Latour has looked to art and museum exhibitions to explore problems, objects, and possible solutions (Latour and Weibel 2020). Art and imaginative thinking are able to disclose new worlds and approaches for new problems, according to Latour. His most recent work of art was an interdisciplinary exhibition that he curated with his long-standing friend and colleague Peter Weibel at the Center for Art and Media Karlsruhe (ZKM) entitled *Critical Zones: Observatories for Earthly Politics* (2020). This exhibition is closely connected to Latour's previously published book *Down to Earth* (2018 [2017]).

In the exhibition that Latour curated with Weibel, diverse works from artists, philosophers, historians, activists, students, and poets were brought together. The exhibition catalogue, which is over 400 pages, has been published as a book by MIT Press under the title *Critical Zones: The Science and Politics of Landing on Earth* (2020). The book is conceived in such a way that readers can study it on its own terms, and it may generate thinking long after the exhibition itself (Latour and Weibel 2020).

In the context of this chapter, I would like to draw attention to three of the contributions to the book, which elegantly and in a hybrid fashion connect some of the thoughts and installations on display in the exhibition. The first contribution I want to highlight was created by Martin Guinard and Bettina Korintenberg, "Observations for Terrestrial Politics: Sensing the Critical Zones" (Latour and Weibel 2020, 402f). It describes some of the curatorial decisions and reflections of the exhibition that finally—and in line with the coproduction theme of the exhibition—was turned into a digital platform due to the COVID-19 pandemic.

In order to connect the visitors of *Critical Zones* with what is at stake for all earthlings and increase knowledge creation through interaction, guests were initially asked a number of critical questions upon entering, that is

> Name something which you depend on and which you have learned is under threat. Why does it touch you? Where is the information from? And which action are you ready to take to slow down the threat? Or, on the contrary, what prevents you from such change? (Latour and Weilbel 2020, 410)

An important dimension of the exposition was to reconnect presence and absence evoked through the phenomena of so-called "ruins and ghosts," and thus to generate increased awareness of the networked structure of matters-of-concern, politics, and behavior.

Inspired by Kenneth Pomeranz's book *The Great Divergence*, the concept "ruin" is connected to both the producer and the product of "ghost acres," which informs the second contribution I want to highlight, Uriel Orlow's multimedia installation "Soil Affinities." In this work, the artist meticulously traces the food supply chain from consumers in Europe to the produce's origin in Africa, "which divide the land one lives on from the [ghost] land one lives off" (Latour and Weibel 2020, 405). "Soil Affinities" refers to the tight interdependence of European and African soil and signifies the radical caesura of lifeworlds associated with two different geographies, the two critical zones.

Inspired by so-called forensic architecture and the endeavor to display "state violence linked to ecological change" (Latour and Weibel 2020, 406), students from the London-based studio ADS7 created "Something in the Air, politics of the Atmosphere," which is the third contribution that I would like to point out. In this work, the artists zoom in on and frame the Chinese Sky River Project in a way that shows how "old geopolitical order of land sovereignty extends to the realm of the atmosphere, transforming it into an aerography where resources and borders are contested" (Latour and Weibel 2020, 407). The Sky River Project, a gigantic geoengineering experiment in China, was carried out to produce artificial rain over an area of 1.6 million square kilometers covering the Tibetan Plateau. Its purpose was to secure water in the Chinese rivers and ameliorate the country's water shortage. The work by ADS7 may thus also be viewed as a contemporary homage to Heidegger's description of the hydroelectric plant, not least because the students provide their installation with the following textual frame: "The atmosphere becomes a space that can be mapped through scientific research, appropriated through technology and administered through both land and geospatial infrastructure" (Latour and Weibel 2020, 407).

In a very concrete yet imaginative sense, the exhibition forces visitors *down to Earth* to the so-called *critical zone*, that is, the thin, fragile, and very diverse layer of soil and life on the surface of the Earth. It is the explicit aim of the exhibition to create the sort of sensitivities to the critical zone that can engender critical and imaginative thoughts and actions and thus help point out encouraging directions for a difficult yet necessary break with the status quo and the (modern) development leading to climate and environmental disaster. Modern humans carry the largest responsibility for the environmental catastrophe of the critical zone. Only "we" have a choice, the power of politics, and the potential to generate an adequate response. In this sense, the asymmetry between humans and nonhumans reappears and can be connected to the criticism developed in chapter 1.

ON THE SHAPING OF LANDING SITES

In his later writings, Latour's views and judgments of modernity come even closer to Heidegger's. This is especially manifest in four specific dimensions of Latour's work: his analysis of the Earth viewed as a globe, his critique of the detrimental modern hunger for energy, his appeal to art, and his fight against human homelessness.

Latour's critique of the global view of the Earth, which was unfolded above, bears clear and interesting resemblances to Heidegger's comprehensive critique in the essay "The Age of the World Picture" (1959). Here, Heidegger unfolds the epistemological underpinnings of the modern sciences, which we also saw Latour criticizing in the section above. Seen from outer space, as Latour pointed out, our planet gives a detached understanding of the Earth, one that neglects the local differences and emphasizes global coherence, which again leads to the production of universal standards that have come to catalyze environmental disasters. Heidegger would support such a critique and add that the distanced view according to which one can talk about such a thing as *a world picture* of a certain epoch is in itself a product of modernity. In Antiquity and the Middle Ages, there was no such abstract world picture, yet in the modern age, it has retrospectively been projected onto these two periods of time and shaped our understanding of them. To see and define a world picture has indeed laid the ground for the emergence of the modern view of the planet that Latour refers to, but seeing the world as a picture "means more than this. We can mean by it the world itself; the totality of beings taken, as it is for us, as standard-giving and obligating" (Heidegger 1998b, 67).

> Where the world becomes picture, beings as a whole are set in place as that for which man is prepared; that which, therefore, he correspondingly intends to bring before him, have before him, and, thereby, in a decisive sense, place before him. (Heidegger 1998b, 67)

As the world turns into a world picture, it is put on display and represented to a subject and at its disposal. Through the world picture, the ground is prepared for manipulation by modern technologies. As Heidegger points out, "The fundamental event of Modernity is the conquest of the world as picture" (Heidegger 1998b, 71).

It is in this context that Heidegger's critique of a modern hydroelectric power plant carries its significance and connects to Latour and Weibel's exhibition of the Chinese rainwater intervention. When the Rhine and the clouds of rain have been analyzed, quantified, and mathematized in terms of modern physics, they increasingly present themselves for the sort of intrusion that changes the river into a source of energy and forces the Rhine and the clouds to deliver resources.

In a world grasped as a world picture, humans are prone to homelessness in the sense that they easily lose their close attachment to a specific place, which is, again, intimately associated with the view of the planet as a gigantic assemblage of resources, which produces estrangement and makes it increasingly difficult to find a place to dwell and cohabit.

In an attempt to (re)establish an attachment to place, architecture—a hybrid of technology and art—may come to play a significant role. Here, in the final chapter of the book, I would like to point out how architecture can get us *down to Earth*, shape a *landing site*, and thus reconnect fundamental insights of Heidegger and Latour in a constructive manner. At the same time, this reflection draws on the previous chapters and shapes the ending of the book.[8]

Heidegger's notion of a human being, Dasein, has at its very core a connection to the surrounding world, as we have seen in previous chapters: Dasein is down to Earth in its core and is defined as being-in-the-world.

> What does being-in mean? [. . .] being-in designates a constitution of being of Dasein and is an *existential*. But we cannot understand by this the objective presence of a material thing (the human body) "in" a being objectively present [. . .]. 'In' stems from *innan*-, to live, *habitare*, to dwell. "An" means I am used to, familiar with, I take care of something. It has the meaning of *colo* in the sense of *habito* and *diligo*. We characterized this being to whom being-in belongs in this meaning as the being which I myself always am. The expression "bin" is connected with *"bei"*. *"Ich bin"* (I am) means I dwell, I stay near . . . the world as something familiar in such and such a way. (Heidegger 2008, 54 f)

In Heidegger's notion of human beings, on one hand, they are directed toward and connected to their local environment; on the other, they dwell and are thus indirectly related to architecture. In Heidegger's text, "Building Dwelling Thinking," he examines the connection between building and dwelling, without explicitly reflecting on the notion of architecture. However, the general drift of his thinking in this text connects very well with the present focus on architecture. The concept of architecture comes from an amalgam of the Greek words *arche* and *techtura*. The former word means *original principle*, and the latter can be traced back to *the practice of weaving*, that is, binding something together. In other words, architecture is a profound and fundamental weaving together. But what is woven together through architecture? Based on the line of thinking above, the answer to this question is humans, artifacts, and the environment. Poetically thought, architecture gathers "the fourfold" of the mortals, the earth, the heavens, and the divinities into one thing: the dwelling place. In successful architectural works, the fourfold unifies and unfolds and becomes inseparable as the mortals *come down to earth* and appreciate the opening of their situated existence as well as their implicit vulnerability.

By reflecting on the significance of architecture, humans may come to understand the basic existential network that sustains their life in a sense that negates the view from nowhere, that is, the view of the planet from outer space, which was opposed by both Latour and Heidegger. Architecture is near and shares fundamental characteristics with *the thing*. How this sort of reflection belongs to architecture is manifest in Heidegger's famous description of a nonmodern farmhouse in the Black Forest. The intrinsic network of humans, artifacts, and natural objects, and processes combined with religious relics, underlines the close family resemblance between Heidegger's fourfold dwelling place and Latour's variegated and far-reaching networks. I quote Heidegger's complete description of the Black Forest farmhouse so that we can appreciate the many more or less evident connections between Latour and Heidegger.

> Let us think for a while of a farmhouse in the Black Forest, which was built some two hundred years ago by the dwelling peasants. Here the self-sufficiency of the power to let earth and sky, the divinities and mortals enter in simple oneness into things ordered the house. It placed the farm on the wind-sheltered mountain slope, looking south, among the meadows close to the spring. It gave it the wide overhanging shingle roof whose proper slope bears up under the burden of snow, and that, reaching deep down, shields the chambers against the storms of the long winter nights. It did not forget the alter corner behind the community table; it made room in its chamber for the hallowed places of childbed and the "tree of the dead"—for that is what they call a coffin there: the totenbaum. And in this way it designed for the different generations under one roof the character of their journey through time. A craft that, itself sprung from dwelling, still uses its tools and its gear as things, built the farmhouse. (Heidegger 1994a, 361f)

Consonant with the farming terminology described by Heidegger, which is part of the endeavor to get down to Earth, Latour adds, "Targeting emancipation through weightlessness does not require the same virtues as targeting emancipation through a process of *plowing*, a way to dig in" (Latour 2018, 81; emphasis added). And Latour continues, "In the latter system, not only points of view, but also points of life proliferate" (Latour 2018, 88).

Based on this comparative interpretation of Latour and Heidegger, I would like to encourage debate and offer a possible answer to the question raised by Latour, "So, the question of what it means for a people to live in space, on land, from a soil is wide open all over again" (Latour and Weibel 2020, 6). To live on land and from the soil has to do with architecture in a fundamental sense. This sort of after-modern inhabiting of the land knows of no absolute barriers between interior and exterior; it is grounded in a place, and shows dependency, which "comes in first of all to limit, then to complicate, then to reconsider the project of emancipation, in order to finally amplify it" (Latour 2018, 83). Such an architecture is fragile, humble, flexible, and in close conversation with its surrounding environment. It is intimately attached to its environment in such a way that a distortion of it also creates a distortion of the architecture in question. It thrives in the richness of the exterior, makes use of materials in a minimalist way, lets them shine, and makes them easy to recycle. Such an architecture weaves together and gathers the fourfold in one dwelling place as it lets the mortals be grounded on earth, with a respectful opening to the great exterior as it generates humility toward and gratitude for the environment.

This kind of architecture stands in sharp contrast to the effective and rationalist conceptions of modernity, which Heidegger objected to and which Latour now hangs out to dry in a very pointed fashion:

> How could we speak of "effectiveness" with respect to technological systems that have not managed to integrate into their design a way to last more than a few decades? How could we call "rationalist" an ideal of civilization guilty of forecasting error so massive that it prevents parents from leaving an inhabited world to their children? (Latour 2017, 66)

One example of an attempt to weave architecture into the environment is the famous "Fallingwater" by Frank Lloyd Wright. This architectural work of art is hidden in the woods of Pennsylvania and exudes respect for its natural surroundings. The house stands in balanced equilibrium with the adjacent forest and enhances the appearance of the falling water below. It does not sap the energy from the water as the hydroelectric plant does but pays tribute to the natural flow of water and embraces it in a poetic fashion. However, the use of reinforced concrete and the size of the single-family home make it unsustainable. It does not provide all the answers to how we may create sustainable architecture on Earth for billions of people. The inspiration is on a conceptual and poetic level. There is a clear and present need for a new generation of architects who use recycled materials and the ideas of Latour and Heidegger to build minimalist refuges or macro-scale collective housing and cityscapes with the capabilities of ANT-hills.

With these converging reflections on architecture as a protective artifact against the perils of modernity, the present exploration with Heidegger and Latour draws to a close. Although Heidegger and Latour have passed away, their legacies endure as guides for the future, offering insights to help their readers navigate the significant challenges that lie ahead. Throughout this book, critical groundwork has been laid for further philosophical investigations of these two intellectual giants, numerous new associations have been uncovered, and connections between them emphasized. I elucidated these various elements and demonstrated how Latour and Heidegger ultimately converge in their warnings against the pitfalls of modernity. This comparative exposition aimed to amplify the impact of their exceptional thinking. While they may not provide all the answers needed in these revolutionary times, they pose vital questions and stimulate further contemplation on some of the most urgent issues of the human era—a time that may be drawing to a close if modern civilization fails to heed the critical insights they offer.

Modernity, technology, and human existence are intricately intertwined. Through the experimental theoretical fieldwork undertaken in this book, these concepts have been examined in various configurations, revealing a landscape fraught with profound dangers that threaten all life on Earth. Yet, our current era is characterized by radical ambivalence, harboring the potential for divergent paths into the future. Modernity appears to be reaching its denouement, which could manifest as either apocalyptic collapse or a truly revolutionary upheaval that challenges the prevailing notion of humans and the Earth as mere resources. Regardless of the trajectory taken, the twenty-first century will profoundly shape the fate of our species and all life on Earth. Heidegger invoked the poet Hölderlin, stating, "But where the danger is, grows // The saving power also" (Heidegger 1977a, 340). Whether this holds true remains to be seen. Judging by contemporary developments, the growth of the saving power appears to lag behind the expansion and intensification of the dangers we face.

NOTES

1. This is also supported by statements of Latour, such as,

Contrary to what makes Heideggerians weep, there is an extraordinary continuity, which historians and philosophers of technology have increasingly made legible, between nuclear plants, missile-guidance systems, computer-chip design, or subway automation and the ancient mixture of society, symbols, and matter that ethnographers and archaeologists have studied for generations in the cultures of New Guinea, Old England, or sixteenth-century Burgundy. (Latour 1999, 195)

2. "No radical revolution can separate us from these pasts, so there is no need for reactionary counter revolutions to lead us back to what has never been abandoned" (Latour 1993, 68).

3. Should we say: At a more precise historical scale, however, the inherent ubiquity or duplicity of the modernizers is not so recent a phenomenon. It was long in coming. Should we choose 1610, 1789, 1945? It does not matter much. (Latour and Weibel 2020, 4)

4. Cf.,

On page four of my daily newspaper, I learn that the measurements taken above the Antarctic are not good this year: the hole in the ozone layer is growing ominously larger. Reading on, I turn from upper-atmosphere chemists to Chief Executive Officers of Atochem and Monsanto, companies that are modifying their assembly lines in order to replace the innocent chlorofluorocarbons, accused of crimes against the ecosphere. A few paragraphs later, I come across heads of state of major industrialized countries who are getting involved with chemistry, refrigerators, aerosols and inert gases. But at the end of the article, I discover that the meteorologists don't agree with chemists; they're talking about cyclical fluctuations unrelated to human activity. So now the industrialists don't know what to do. The heads of state are also holding back. Should we wait? Is it already too late? Toward the bottom of the page, Third World countries and ecologists add their grain of salt and talk about international treaties, moratoriums, the rights of future generations, and the right to development. (Latour, 1993, 1)

Confer also the book that I coedited with Pernille Almlund and Per Homann Jespersen (Almlund et al., 2012).

5. In times of pervasive technology, Latour should be asked if he has been fighting the right enemy. The following questions, which Latour himself has posed , should therefore be emphasized and, in the context of the controversy between Latour and Heidegger, readdressed:

Would it not be rather terrible if we were still training young kids—yes, young recruits, young cadets—for wars that are no longer possible, fighting enemies long gone, conquering territories that no longer exist, leaving them ill-equipped in the face of threats we had not anticipated, for which we are so thoroughly unprepared? Generals have always been accused of being on the ready one war late—especially French generals, especially these days. Would it be surprising, after all, if intellectuals were also one war late, one critique late—especially French intellectuals, especially now? (Latour 2004, 225f). Cf. also Zuboff (2019)

6. Here it is also interesting to recall Latour's former critique of Heidegger, which suggests that the late Latour has changed his views: "As Lévi-Strauss says 'the barbarian is first and foremost the man who believes in barbarism'" (Latour 1993, 66).

7. This is the sort of rupture in history to which Latour now gives his full attention.

8. Cf. Riis (2013, 2021).

Bibliography

Almlund, P. et al. (2012). *Rethinking Climate Change Research: Clean-technology, Culture and Communication.* Ashgate.

Anderson, J.C., Clarke, E.J., Arkin, A.P., & Voigt, C.A. (2005). "Environmentally Controlled Invasion of Cancer Cells by Engineered Bacteria." *Journal of Molecular Biology*, 355, pp. 619–627.

Aristotle. (1984a). "Physics." In *The Complete Works of Aristotle*, pp. 315–446. Princeton University Press.

Aristotle. (1984b). "Politics." In *The Complete Works of Aristotle*, pp. 1986–2129. Princeton University Press.

Balslev, J. (2018). *Kritik af den digitale fornuft - i uddannelse.* Hogrefe Psykologisk Forlag

Banksy. (2014). "Goodreads." http://www.goodreads.com/author/quotes/28811.Banksy (Verified 20.02.2022).

Bennett, J. (2010). *Vibrant Matter: A Political Ecology of Things.* Duke University Press.

Blok, A., & Jensen, T.E. (2012). *Bruno Latour: Hybrid Thoughts in a Hybrid World.* Routledge.

Bloor, David. (1999). "Anti-Latour." *Studies in History and Philosophy of Science*, 30(1), pp. 81–112.

Bruno, G. (1977 [1584]). *Von der Ursache, dem Prinzip und dem Einen.* Felix Meiner Verlag.

Bultmann, R. (2009). *Bultmann, Rudolf / Martin Heidegger: Briefwechsel 1925 bis 1975.* Klostermann Verlag.

Carr, N. (2008). "Is Google Making Us Stupid? What the Internet Is Doing to Our Brains." *Atlantic.* http://www.theatlantic.com/magazine/archive/2008/07/is-google-making-us-stupid/6868/ (Verified 25.02.2022)

Dictionary. (2013). "Dictionary Reference." http://dictionary.reference.com/browse/relent (Verified 01.03.2022)

Dryfus, H., & Spinosa, C. (1997). "Highway Bridges and Feasts: Heidegger and Borgmann on How to Affirm Technology." *Man and World*, 30, pp. 159–177.

Dryfus, H., & Spinosa, C. (1999). "Coping with Things-in-Themselves: A Practice-Based Phenomenological Argument for Realism." *Inquiry*, 24, pp. 49–78.

Duden. (1997). *Das Herkunftswörterbuch*. Duden.

Feenberg, A. (1995). *Alternative Modernity: The Technical Turn in Philosophy and Social Theory*. University of California Press.

Feenberg, A. (2005). *Heidegger and Marcuse: The Catastrophe and Redemption of History*. Routledge.

Finley, M. I. (1980). *Ancient Slavery and Modern Ideology*. Chatto and Windus.

Franklin, S. (2007). *Dolly Mixtures: The Remaking of Genealogy*. Duke University Press Books.

Galison, P. (1994). "The Ontology of the Enemy: Norbert Wiener and the Cybernetic Vision." *Critical Inquiry*, 21, pp. 228–266.

Galison, P. (2006). "Breakdown." In Selinger, E. (Ed.), *Postphenomenology: A Critical Companion to Ihde*. SUNY Press.

Glazebrook, T. (2000). *Heidegger's Philosophy of Science*. Fordham University Press.

Gill, B. (2008). "Über Whitehead und Mead zur Akteur-Netzwerk-Theorie: Die Überwindung des Dualismus von Geist und Materie – und der Preis, der dafür zu zahlen ist." In Kneer, G., Schroer, M., & Schüttpelz, E. (Eds.), *Bruno Latours Kollektiv: Kontroversen zur Entgrenzung des Sozialen*, pp. 47–75. Suhrkamp Verlag.

Grondin, J. (2001). "Die Wiedererweckung der Seinsfrage auf dem Weg einer phänomenologisch-hermeneutischen Destruktion." In Rentsch, T. (Ed.), *Klassiker Auslegen: Sein und Zeit*, pp. 1–27. Akademie Verlag GmbH.

Gursky, A. (1999). "Rhine II." https://www.tate.org.uk/art/artworks/gursky-the-rhine-ii-p78372 (Verified 25.02.2024).

Guynn, J. (2012). "Steve Jobs Fans Continue to Make Pilgrimage to His Palo Alto Home." *Los Angeles Times*. http://www.latimes.com/business/technology/la-fi-tn-steve-jobs-house-20120217 (Verified 25.02.2022).

Harman, G. (2002). *Tool-Being: Heidegger and the Metaphysics of Objects*. Open Court Publishing.

Harman, G. (2005). "Heidegger on Objects and Things." In Latour, B., & Weibel, P. (Eds.), *Making Things Public: Atmospheres of Democracy*, pp. 268–271. MIT Press.

Harman, G. (2009). *Prince of Networks: Bruno Latour and Metaphysics*. Re.press.

Harman, G. (2011). *The Quadruple Object*. Zero Books.

Heidegger, M. (1966 [1959]). *Discourse on Thinking*. Harper & Row.

Heidegger, M. (1977a [1954]). "The Question Concerning Technology." In Krell, D.F. (Ed.), *Basic Writings*, pp. 311–341. Harper & Row.

Heidegger, M. (1977b [1950]). "The Origin of the Work of Art." In Krell, D.F. (Ed.), *Basic Writings*, pp. 143–212. Harper & Row.

Heidegger, M. (1977c [1950]). "The End of Philosophy and the Task of Thinking." In Krell, D.F. (Ed.), *Basic Writings*, pp. 431–449. Harper & Row.

Heidegger, M. (1977d [1967]). "Letter on Humanism." In Krell, D.F. (Ed.), *Basic Writings*, pp. 217–265. Harper & Row.
Heidegger, M. (1977f [1954]). "Building Dwelling Thinking." In Krell, D.F. (Ed.), *Basic Writings*, pp. 347–363. Harper & Row.
Heidegger, M. (1996 [1927]). *Being and Time*. New York> SUNY Press.
Heidegger, M. (1998a [1967]). *Pathmarks*. Cambridge University Press.
Heidegger, M. (1998b [1967]). "On the Essence and Concept of ΦΥΣΙΣ in Aristotle's Physics B, I." In *Pathmarks*, pp. 183–230. Cambridge University Press.
Heidegger, M. (1999b [1989]). *Contributions to Philosophy: From Enowning* (Emad, P. & Maly, K., Trans.). Indiana University Press.
Heidegger, M. (2000b [1981]). *Elucidations of Hölderlin's Poetry*. Humanity Books.
Heidegger, M. (2001 [1954]). "The Thing." In *Poetry, Language, and Thought*, pp. 161–180. Harper and Row.
Heidegger, M. (2002a [1985]). *Phenomenological Interpretations of Aristotle: Initiation into Phenomenological Research* (Rojcenricz, R., Trans.). Indiana University Press.
Heidegger, M. (2002b). *Supplements: From the Earliest Essays to Being and Time and Beyond* (van Buren, J., Ed.). SUNY Press.
Heidegger, M. (2003). "Overcoming Metaphysics." In Stambaugh, J. (Trans.), *The End of Philosophy*, pp. 84–110. University of Chicago Press.
Heidegger, M. (2008 [1927]). *Being and Time*. New York: HarperCollins.
Hilt, A. (2005). "Die Frage nach dem Menschen: Anthropologische Philosophie bei Helmuth Plessner und Martin Heidegger." In Figal, G. (Ed.), *Internationales Jahrbuch für Hermeneutik*, 4, pp. 275–320.
Hui, Y. (2016). *The Question Concerning Technology In China: An Essay in Cosmotechnics*. Urbanomic Media Ltd.
Husserl, E. (2001 [1900/1901]). *Logical Investigations* (Moran, D., Ed., 2nd ed.). 2 vols. Routledge.
Ihde, D. (1979). *Technics and Praxis*. D. Reidel Publishing Company.
Ihde, D. (1995). *Postphenomenology: Essays in the Postmodern Context*. Northwestern University Press.
Ihde, D. (2003) "If Phenomenology Is an Albatross, Is *Post-phenomenology* Possible?" In Ihde, D. & Selinger, E. (Eds.), *Chasing Technoscience: Matrix for Materiality*, pp. 131–146. Indiana University Press.
Ihde, D. (2008). *Ironic Technics*. Automatic Press / VIP.
Ihde, D. (2010). *Heidegger's Technologies: Postphenomenological Perspectives*. Fordham University Press.
Ihde, D., & Selinger, E. (Eds.). (2003). *Chasing Technoscience: Matrix for Materiality*. Indiana University Press.
Jasanoff, S. (2005). *Designs on Nature*. Princeton University Press.
Jasanoff, S., & Kim, S. (2015). *Dreamscapes of Modernity: Sociotechnical Imaginaries and the Fabrication of Power*. University of Chicago Press.

Jensen, C.B., & Gad, C. (2013). "Spørgsmål til teknologierne: Om komplekser af ting og tænkning." In Schiølin, K., & Riis, S. (Eds.), *Nye spørgsmål om teknikken*, pp. 187–207. Aarhus Universitetsforlag.

Jünger, E. (1982). *Der Arbeiter*. Klett-Cotta Verlag.

Kalcyk, H. (1982). *Untersuchungen zum Attischen Silberbergbau: Gebietstruktur, Geschichte und Technik*. Verlag Peter Lang.

Khong, L. (2003). "Actants and Enframing: Heidegger and Latour on Technology." *Studies in History and Philosophy of Science*, 34, pp. 693–704.

Kingsley, D., & Urry, J. (2010). *After the Car*. Polity Press.

Kisiel, T. (1993). *The Genesis of Heidegger's Being and Time*. University of California Press.

Kneer, G., Schroer, M., & Schüttpelz, E. (Eds.). (2008). *Bruno Latours Kollektiv: Kontroversen zur Entgrenzung des Sozialen*, pp. 47–75. Suhrkamp Verlag.

Kochan, J. (2010). "Latour's Heidegger." *Social Studies of Science*, 40(4), pp. 579–598.

Kuhn, T. (1970). *The Structure of Scientific Revolutions* (2nd ed.). University of Chicago Press.

Landecker, H. (2007). *Culturing Life: How Cells Became Technologies*. Harvard University Press.

Latour, B. (1986 [1976]). *Laboratory Life: The Construction of Scientific Facts*. Princeton University Press.

Latour, B. (1993a [1991]). *We Have Never Been Modern*. Harvard University Press.

Latour, B. (1993b [1984]). *The Pasteurization of France*. Harvard University Press.

Latour, B. (1999). *Pandora's Hope: Essays on the Reality of Science Studies*. Harvard University Press.

Latour, B. (2002b). "The Science Wars: A Dialogue." *Common Knowledge*, 8(1), pp. 71–79.

Latour, B. (2002c [1992]). *Aramis or the Love of Technology*. Harvard University Press.

Latour, B. (2003). "The Promise of Constructivism." In Ihde, D., & Selinger, E. (Eds.), *Chasing Technoscience: Matrix for Materiality*, pp. 27–46. Indiana University Press.

Latour, B. (2004a). "Why Has Critique Run out of Steam? From Matters of Fact to Matters of Concern." *Critical Inquiry*, 30, pp. 225–248.

Latour, B. (2004b). "A Dialog on Actor Network Theory with a (Somewhat) Socratic Professor." In Avgerou, C., Ciborra, C., & Land, F.F. (Eds.), *The Social Study of Information and Communication Study*, pp. 62–76. Oxford University Press.

Latour, B. (2005b). *Reassembling the Social: An Introduction to Actor-Network-Theory*. Oxford University Press.

Latour, B. (2008b). *What is the Style of Matters of Concern?* Royal Van Gorcum.

Latour, B. (2010b [1995]). *On the Modern Cult of the Factish Gods*. Duke University Press.

Latour, B. (2010c). "Coming Out as a Philosopher." *Social Studies of Science*, 40(4), pp. 599–608.

Latour, B. (2011). "Reflections on Etienne Souriau's Les différents modes d'existence." In Bryant, L.R., Srnicek, N., & Harman, G. (Eds.), *The Speculative Turn: Continental Materialism and Realism*, pp. 304–333. Re.press.

Latour, B. (2013). *An Inquiry into Modes of Existence; An Anthropology of the Moderns*. Harvard University Press.

Latour, B. (2017). *Facing Gaia: Eight Lectures on the New Climatic Regime*. Polity Press.

Latour, B. (2018). *Down to Earth: Politics in the New Climatic Regime*. Polity Press.

Latour, B., & Weibel, P. (2020). *Critical Zones: The Science and Politics of Landing on Earth*. MIT Press.

Lucker, A. (2007). "Dinge–Zeuge–Werke: Technik und Kunst bei Heidegger." In Hubig, C., Luckner, A., & Mazouz, N. (Eds.), *Handeln und Technik – mit und ohne Heidegger*, pp. 193–210. LIT Verlag.

Magee, J. G. (2013 [1941]). "High Flight." http://www.arlingtoncemetery.net/highflig.htm (Verified 01.03.2013)

Maniglier, P. (2013). "Stephen Muecke on Bruno Latour's Modes of Existence." *Le Monde*. http://urbanchoreography.net/2013/01/14/stephen-muecke-on-bruno-latours-modes-of-existence/ (Verified 14.03.2022)

Markoff, J., & Sengupta, S. (2011). "Separating You and Me? 4.74 Degrees." *New York Times*. https://www.nytimes.com/2011/11/22/technology/between-you-and-me-4-74-degrees.html (Verified 25.02.2022)

Marinitte, F. T. (2013 [1909]). "Futurist Manifesto." http://vserver1.cscs.lsa.umich.edu/~crshalizi/T4PM/ futurist-manifesto.html (Verified 01.03.2022)

Melson, A., & Riis, S. (2015). "The Living-Dead and the Existence of God." *Danish Yearbook of Philosophy*, 47, pp. 65–86.

Milgram, S. (1967). "The Small World Problem." *Psychology Today*, 2, pp. 60–67.

Munk, A. K., & Abrahamsson, S. (2012). "Empiricist Interventions: Strategy and Tactics on the Ontopolitical Battlefield." *Science Studies*, 25, pp. 52–70.

Nietzsche, F. (1956 [1887]). *The Birth of Tragedy & The Genealogy of Morals*. Doubleday Publishing.

Nitsch, W. (2008). "Dädalus und Aramis: Latours symmetrische Anthropologie der Technik." In Kneer, G., Schroer, M, & Schüttpelz, E. (Eds.), *Bruno Latours Kollektiv: Kontroversen zur Entgrenzung des Sozialen*, pp. 219–233. Suhrkamp Verlag.

Nordmann, A. (2008). "Technology Naturalized: A Challenge to Design for the Human Scale." In Vermaas, P.E., Kroes, P., Light, A., & Moore, S.A. (Eds.), *Philosophy and Design: From Engineering to Architecture*, pp. 173–184. Springer.

Ott, H. (1992). *Martin Heidegger: Unterwegs zu seiner Biographie*. Campus Verlag.

Pickering, A. (2002). "Cybernetics and the Mangle: Ashby, Beer and Pask." *Social Studies of Science*, 32, pp. 413–437.

Pickering, A. (2009). "Beyond Design: Cybernetics, Biological Computers and Hylozoism." *Synthese*, 168, pp. 469–491.

Platte, T. (2004). *Die Konstellation des Übergangs: Technik und Würde bei Heidegger*. Duncker & Humblot.

Rannard, G. (2022). "Three scientists win Nobel for chemistry 'Lego'". BBC. https://www.bbc.com/news/science-environment-63121338 (verified 11.11).

Riis, S. (2008). "The Symmetry between Bruno Latour and Martin Heidegger: the Technique of Turning a Police Officer into a Speed Bump." *Social Studies of Science*, 38(2), pp. 285–301.
Riis, S. (2013). "Reframing Architecture." *Foundations of Science*, 18(1), pp. 205–211.
Riis, S. (2015). "A Century on Speed: Reflections on Movement and Mobility in the Twentieth Century." In Rosenberger, R., & Verbeek, P-P. (Eds.), *Postphenomenological Investigations: Essays on Human-Technology Relations*, pp. 159–174. Lexington Books.
Riis, S. (2017). "ICT Literacy: An Imperative of the Twenty-First Century." *Foundations of Science*, 22(2), pp. 385–394.
Riis, S. (2018). *Unframing Martin Heidegger's Understanding of Technology: On the Essential Connection between Technology, Art, and History*. Lexington Books.
Riis, S. (2021). "Building Dwelling and the End of Thinking." In Botin, L., & Hyams, I. B. (Eds.), *Postphenomenology and Architecture: Human Technology Relations in the Built Environment*, pp. 213–227. Lexington Books.
Rojcewicz, R. (2006). *The Gods and Technology*. SUNY Press.
Rosenblueth, A., & Wiener, N. (1950). "Purposeful and Non-Purposeful Behavior." *Philosophy of Science*, 17, pp. 318–326.
Ruin, H. (2010). "Ge-stell: Enframing as the Essence of Technology." In Davis, B.W. (Ed.), *Martin Heidegger: Key Concepts*, pp. 183–194. Routledge.
Savage, D.F., Way, J., & Silver, P.A. (2008). "Defossiling Fuel: How Synthetic Biology Can Transform Biofuel Production." *ACS Chemical Biology*, 3(1), pp. 13–16.
Scharff, R.C., & Dusek, V. (2003). "Introduction to Part IV: Heidegger on Technology." In Scharff, R.C., & Dusek, V. (Eds.), *Philosophy of Technology: The Technological Condition: An Anthology*, pp. 247–251. Blackwell.
Schøilin, K. (2012). "Follow the Verbs! – A Contribution to the Study of the Heidegger–Latour Connection." *Social Studies of Science*, 42(5), pp. 776–787.
Smith, A. (2010). *Wealth of Nations*. Capstone.
Spinoza, B.d. (1955). *Ethics*. In *The Chief Works of Benedict de Spinoza*. Dover Publications.
Tasheva, G. (2001). "Zeit und Differenz: Soziologie Jenseits von Ontologie." In Weiß, J. (Ed.), *Die Jemeiningkeit des Mitseins: Die Daseinsanalytik Martin Heideggers und die Kritik der soziologischen Vernunft*, pp. 149–173. UVK Verlag.
Tasheva, G. (2009). "Tod Opferritual, Theatralisierung. Spaltungen am Ursprung der Gesellschaft." In Willems, H. (Ed.), *Theatralisierung der Gesellschaft 1: Soziologische Theorie und Zeitdiagnose*, pp. 279–301. VS Verlag.
Thomä, D. (2003). *Heidegger Handbuch*. Verlag J. B. Metzler.
Verbeek, P. (2005). *What Things Do*. The Penn State Press.
Verbeek, P. (2016). "Toward a Theory of Technological Mediation: A Program for Postphenomenological Research." In Berg, J.K., Friis, O., & Crease, R.C. (Eds.), *Technoscience and Postphenomenology: The Manhattan Papers*, pp. 189–204. Lexington Books.

Verbeek, P., & Rosenberger, R. (2015). "A Field Guide for Postphenomenology." In Rosenberger, R., & Verbeek, P. (Eds.), *Postphenomenological Investigations: Essays on Human-Technology Relations*, pp. 9–41. Lexington Books.

Wade, N. (2010). "Researchers Say They Created a Synthetic Cell." *New York Times*. http://www.nytimes.com/2010/05/21/science/21cell.html?hpw (Verified 11.11)

Wamberg, J. (2009). *Landscape as World Picture: Tracing Cultural Evolution in Images*. University of Aarhus Press.

Weiß, J. (2001). "Einleitung." In Weiß, J. (Ed.), *Die Jemeiningkeit des Mitseins: Die Daseinsanalytik Martin Heideggers und die Kritik der soziologischen Vernunft*, pp. 11–56. UVK Verlag.

Wiener, N. (1948). *Cybernetics or Control and Communication in the Animal and the Machine*. The Technology Press.

Yoxen, E. (1983). *The Gene Business: Who Should Control Biotechnology*. Pan Books.

Zammito, J.H (2004). *A Nice Derangement of Epistemes: Post-positivism in the Study of Science from Quine to Latour*. The University of Chicago Press.

Zimmerman, M. (1990). *Heidegger's Confrontation with Modernity: Technology, Politics and Art*. Indiana University Press.

Zuboff, S. (2019). *The Age of Surveillance Capitalism*. Public Affairs.

Index

actant/actor, 51, 52, 55, 60
Actor-Network Theory (ANT), 43, 44, 46, 51–55, 58–60, 107, 125
airplane, 21–22, 66, 67, 82–86. *See also* John Gillespie Magee, Jr
aletheia, 5, 33, 34, 38, 41, 50
alienation, 16
analogy, 54, 56, 81, 109, 117
ancient, 15, 19, 33, 34, 50, 84, 93, 94, 98, 103, 108. *See also* ancient technology
animism, 62. *See also* pantheism
anthropocentric, 96
anthropology. *See also* Philosophical Anthropology
anxiety, 46, 49, 50, 52. *See also* angst
appearance, 17, 34, 38, 99, 125. *See also* semblance
Aramis, 51, 52. *See also* network
architecture, 117, 121, 123–25. *See also* Frank Lloyd Wright
art/artwork/artist, 9, 11, 25, 26, 35, 37, 38, 40, 41, 43, 51, 121
artifact, 15, 16, 33, 43, 76, 78, 82, 84, 85, 125
Artificial Intelligence (AI), 20
association, 3, 51, 58, 59, 61, 66, 69, 87, 112, 113

asymmetry, 5, 14, 115, 119, 121. *See also* symmetry
authenticity, 6, 60. *See also* Mineness

beginning, 3, 17, 23, 47, 56, 61, 63, 82, 93, 94, 99, 112, 115, 116, 118
Being/Beings, 2, 4, 7, 33–35, 37, 38, 41, 43, 45–50, 59, 63, 67, 69–71, 87, 92–94, 110
being-in-the-world, 43–47, 49, 50, 59, 61, 96, 123. *See also* Dasein
belonging, 10, 14, 70
bifurcation, 54, 91, 102, 103
black box, 15, 16, 59
Bompied, Edmond, 51
bridge, 15, 78, 79
bringing forth or bringing s.th. into appearance, 17, 33–35, 38
Bruno, Giordano, 54

calculating/calculate, 18, 20, 21, 25, 30, 31, 78, 88, 89
car, 66, 76, 84, 89
challenge, 6, 17, 18, 22, 25, 34, 44, 76, 77, 85, 116. *See also* challenging revealing
challenging revealing, 18, 20, 21, 34
civilization, 1, 88, 116, 125, 126

climate change, 1, 76, 115, 116, 118
clearing, 82
collective, 5, 16, 19–27, 43, 47, 51–53, 55, 59–62, 75, 81, 125
combination, 14, 15, 24
command, 18, 22
computer, 14, 19, 20, 52, 88, 126
concealment/concealing, 38, 41, 112
concept, 1, 5, 6, 8, 14, 16–18, 20, 21, 23, 29, 31, 36, 39, 43, 45, 50, 52, 54, 57–62, 66, 69, 71–75, 77–83, 86, 91, 94, 96, 98, 99, 101, 102, 104, 106, 107, 109, 118, 119, 121, 123
connection, 4, 24, 30, 33–37, 49, 57, 58, 71, 73, 77, 80, 92, 97, 98, 103, 104, 119, 123
continuity/discontinuity, 52, 53, 55, 57, 65, 74, 88, 107, 115, 116
control, 1, 13–15, 18, 20, 23, 27
controversy, 2, 42, 127
correctness, 33, 68, 80, 118
correspondence, 3, 34, 99, 104. *See also* truth
craftsman, 43, 83, 124
critique, 3, 6–8, 14–16, 25, 29, 30, 32–37, 43, 44, 55, 65, 67, 72, 81, 89, 101–4, 122, 127
crossing the line, 15, 16, 24. *See also* technical mediation
culture, 31, 56, 73, 76, 84, 88, 98, 103
curiosity, 49, 60
cybernetics, 63

Daedalus, 15, 19–22, 24. *See also* monsters
danger, 1, 5–7, 22, 24, 25, 29, 31, 33, 34, 37, 40, 42, 62, 66, 67, 74, 75, 78, 85, 88, 116, 118, 126. *See also* supreme danger
Dasein, 4, 35, 36, 43–50, 59–62, 87, 96–98, 100, 101, 123
death, 1, 6, 8, 43, 44, 46–48, 53–57, 60, 61, 63, 80, 83, 115, 118
delegation, 22
demands, 15, 17–19. *See also* challenge

destining/destiny, 14, 24, 36, 86, 88
disclose, 26, 29, 39, 49, 50, 63, 71, 78, 81, 83, 95, 105, 120
divine/divinities, 24, 31, 54, 59, 70, 71, 78, 80, 81, 83, 84, 89, 123, 124
double-click, 9
dualism, 26
dwelling, 36, 72, 77–79, 101, 104, 123–25

Earth, 1, 25, 35, 38, 63, 70, 75, 79, 81–87, 89, 116–26
energy, 17, 18, 34, 53, 54, 122, 123, 125
enframing, 6, 16, 17, 19–25, 33, 34, 41, 66, 88. *See also* Essence
engineering/engineer, 15, 20–22, 24, 39, 52. *See also* Daedalus
Enlightenment, 31, 32, 38. *See also* Scientific Revolution
epistemology, 73
epoch, 4, 37, 41, 78, 88, 115, 119, 122. *See also* Modernity
equipment, 19, 33, 35, 59, 62, 70–72, 79. *See also* Ready-to-hand
essence, 16, 17, 19–24, 29, 34, 35, 48, 66, 79, 84, 85, 88. *See also* Essence of technology
ethics, 5, 21, 54, 62. *See also* Values
event, 35, 36, 41, 67, 69, 79, 122. *See also* Truth
existence, 7, 35, 38, 41, 43, 44, 48–50, 53, 55, 57, 60–63, 73, 74, 81, 83, 84, 86, 91, 92, 101, 103, 104, 107–9
experience, 5, 8, 34, 46, 47, 49, 50, 55, 60, 68, 70, 71, 73, 75, 79, 80, 82–85, 97–100, 102–5, 107–10, 113. *See also* Death
experiment, 1, 2, 8–11, 104, 121, 126

fact, 2, 5–7, 10, 13, 21, 24, 27, 37, 45, 48, 52, 55, 56, 61, 66, 68, 73, 75, 76, 80, 86, 91–94, 96, 98, 101, 105, 116, 119
factical, 57, 60, 97
fantastical, 57

fetish, 56, 57
fieldwork, 8–11, 19, 25, 29, 92, 126
finitude, 4, 44, 46, 80
folding, 14, 24
founding/foundation, 9, 30, 35, 43, 54, 58, 63, 95, 99, 101, 111
framework, 5, 6, 8, 10, 16, 21, 24, 33, 35, 39, 45, 80, 82, 83, 85, 87, 89, 95, 100, 101, 105, 107–11, 115
free/freedom, 14, 21, 22, 31, 34, 41, 62, 76, 83, 110–13
fourfold, 7, 66, 70–73, 78–89, 123–25
function, 6, 9, 12, 23, 27, 33, 35, 45–47, 50, 51, 53, 59, 61, 62, 69, 73, 84, 99, 104
future, 32, 38–40, 46, 92, 125–27

gathering, 65, 72–74, 79, 80, 82, 83, 85–87. *See also* fourfold
Genealogy, 9, 97, 98
global, 1, 23, 52, 53, 76, 86, 115, 116, 122
goal, 18, 19, 27, 32, 51, 64, 100, 107, 108, 111
god(s), 30, 31, 54, 55, 57–59, 73, 83, 84, 87–89
granting, 4
Greek temple, 5, 33, 34, 36, 50, 98, 108, 123. *See also* artwork
grounding, 1, 2, 17, 22, 50, 82, 83, 92, 117, 122, 125
growth, 118, 126

happening, 20, 25, 49, 68, 102, 106, 116
hermeneutic, 3, 9, 10, 66, 77, 79, 81, 82, 84, 85, 98, 100–102, 105, 110
history, 3–5, 10, 30, 75, 88, 92, 95–98, 100–103, 111, 115, 127
home/homelessness, 67, 75, 85, 86, 98, 119, 122, 123, 125
human being/Humans, 3–6, 13–15, 19–23, 31, 34, 35, 50, 59, 61, 62, 65, 95, 96, 101, 104, 105, 110, 123
humanism, 33, 35, 36, 41
hybridization/hybrids, 14, 30–32, 40, 52

hydroelectric plant, 18, 30, 31, 33, 34, 89, 121, 125. *See also* Rhine

idea, 22, 25, 31, 32, 35, 38, 44, 47, 51, 54, 55, 57, 62, 66, 76, 80, 81, 86, 93–95, 103, 105, 112, 118, 125
image, 25, 39, 47, 51
immanence, 55–57, 59
individuality, 6, 23, 44, 52, 56, 57, 59, 61, 62
innovation, 49, 57, 67, 68, 85
intentionality, 22, 23, 52, 59, 105
interference, 14, 24
interpretation, 2, 6, 8, 10, 14, 25, 32, 36, 37, 39, 40, 43, 45–47, 51, 55, 57, 60, 66, 72, 73, 77, 79–84, 87, 89, 92, 97, 98, 100, 104, 109–11, 124
intrinsic value, 3, 10, 34, 124
involvement (*Bewandtnis*), 45, 102, 108

jug, 68–72, 74, 84, 86

Kant, 72, 73, 87

landing site, 122–26
language, 8, 9, 15, 20, 37, 56, 95, 101, 103, 108, 113
lifeworld, 5, 10, 45, 69, 71, 102, 104, 109, 121
logic, 19, 38, 55, 71, 94, 95
logistics, 52
logos, 98–100, 103, 108, 110

machine, 11, 14, 18, 21, 22, 52, 61, 63
Magee, John, Gillespie Jr., 83. *See also* Airplane
making, 41, 55, 69, 79, 112
mathematics, 69, 112
matters of concern, 4, 73, 75, 76, 80
matters of fact, 73, 75, 80
meaning, 5, 7, 13, 15, 18, 21, 23, 24, 27, 29, 31, 45, 46, 48, 49, 56, 58, 59, 61, 64, 67, 71–73, 77, 79–81, 83–86, 89, 93–96, 98–100, 102–6, 110, 111, 123
measure, 39, 69, 72, 87, 104, 127

mediation, 6, 13–21, 23–25, 27, 32, 40, 41, 43, 77, 84, 109, 113
metaphysics, 11, 30, 94, 118
method(s), 9, 63, 91–93, 95, 96, 113
Middle Kingdom, 109
mineness, 48, 49, 60, 63
mobilization, 20, 22, 23
modern/modernity, 1–4, 6–9, 11, 13, 16, 17, 20, 22–25, 29–35, 37–41, 43, 52, 55, 56, 66–68, 73, 74, 76–80, 82–86, 88, 89, 98, 101, 102, 104, 106, 107, 112, 115–19, 121, 122, 125, 126
modern Constitution, 30–32, 37–41, 43
Modes of Existence, 7, 44, 61, 76, 91, 92, 101, 103, 107, 108
monsters, 6, 13–27, 32
moral, 5
mortals, the, 61, 70, 73, 80, 81, 84, 87, 89, 123–25
movement, 31, 32, 44, 53–55, 58, 72, 75, 80, 83, 106, 112, 113

nameless, 35, 36
nature, 3, 17–20, 24, 26, 31, 34–37, 54, 55, 64, 71, 73, 74, 78, 86, 87, 95, 100, 103, 111, 118, 119
nearness, 35, 67–70, 72, 73, 79, 82, 85–87
network, 6, 18, 23, 26, 29, 31, 32, 37–40, 43, 44, 50–61, 64, 71, 72, 75, 76, 78, 80, 84–86, 89, 99, 103, 106–8, 110–12, 121, 124
nonhuman, 3–5, 13–16, 18–26, 31, 32, 41–44, 47, 50–53, 55, 57, 59, 62, 108, 109, 112, 115, 116, 119, 121
nonmodern, 7, 30, 32, 40, 41, 124
nonmodern Constitution, 30–32, 40, 41

object(s), 5, 7, 10, 11, 14, 16, 21, 22, 24, 32–34, 36, 38, 39, 43, 45, 50, 52–54, 56, 59, 60, 63, 67, 69, 74–76, 80, 84, 87, 88, 91, 95, 102–4, 107, 109, 116, 118–20, 123–25
one dimensional, 66, 79, 82, 85

ontology, 15, 19, 23, 30–32, 35, 44, 45, 47, 52, 76, 88, 92–98, 100, 101, 104, 108–11
opening/openness, 26, 35, 37, 66, 79, 81, 82, 94, 124
ordering, 17, 18, 20, 24
origin, 5, 6, 9, 11, 35, 39, 42, 45, 57, 58, 61, 73, 74, 93, 95, 97, 98, 100, 116, 121, 123

pantheism, 53, 54, 56, 62
paradigm, 10, 17, 18, 26, 38, 75, 76, 112. *See also* Paradigm shift
paradigm shift, 17
perfection, 51, 58, 61
phenomenology/phenomenon, 5, 7, 11, 26, 35, 42, 46, 48, 49, 53, 63, 67, 91–93, 96–100, 102–4, 109–13
philosophy, 2–4, 8, 10, 12, 25, 30, 33, 37, 64, 76, 80, 88, 92, 93, 103, 111, 112, 119
place(s), 8, 11, 12, 15, 17, 19, 22, 23, 26, 27, 32, 33, 36, 40, 41, 47, 55, 65–68, 71, 74, 75, 77, 82, 84–87, 99, 101, 104, 106, 110, 111, 116–20, 122–25
planet, 15, 23, 115, 117–19, 122–24
plant, 18, 19, 30, 31, 33, 34, 36, 88, 89, 121, 122, 125, 126
poetic reasoning, 7
poetry, 31
politics, 37, 74–77, 86, 88, 117–19, 121
possibility, 21, 36, 46–49, 55, 60, 61, 69, 78, 79, 95
postmodern/postmodernism, 78
practice, 2, 9, 31, 32, 40, 52, 56, 65, 69–71, 76, 77, 98, 100–110, 118, 123. *See also* Practitioner
practitioner, 102, 107–9
prepositions, 107, 110
presence, 15, 36, 69, 72, 78, 83, 85, 89, 120
present-at-hand, 45, 46, 69, 70, 87
production, 4, 14, 16, 21, 25, 31, 33, 41, 76, 86, 98, 105–9, 122

purification, 30–32, 40, 41

Question, 1, 6, 7, 9, 20, 24, 29, 34, 37, 41, 49, 51, 54, 55, 61, 63, 65–67, 78, 79, 86, 93–96, 98, 100, 104, 115, 117, 119, 120, 123–27

rationality, 57
ready-to-hand, 46, 49, 66–73, 87. *See also* Readiness
realism, 56, 97, 101
reality, 64, 68, 69, 111
relation/relationship, 1, 4, 6, 8, 13, 14, 19, 23–25, 35, 36, 38, 39, 43–46, 48, 50–52, 55, 56, 59, 61, 65, 67, 69–74, 78, 80, 81, 83–86, 94–96, 99–102, 104, 107, 111, 112
relativism, 3, 39, 41, 44, 55, 62, 68, 73, 108, 110
religion, 3, 7, 11, 37, 56–58, 108, 111, 117
representation/representing, 32, 39, 47, 48, 51, 69
research, 3, 5, 8–10, 17, 21, 26, 32, 43, 49, 66, 74, 78, 79, 91–93, 95, 96, 98, 100–102, 107, 108, 111, 121
resources, 18, 20–23, 34, 35, 66, 85, 86, 121, 123, 126. *See also* Standing-reserve
responsibility, 5, 115, 116, 119, 121
revealing, 5, 17, 18, 20, 22, 24–26, 31, 33, 34, 38–43, 66, 111, 116, 126
revolution, 17, 30, 32, 36, 47, 67, 72, 75, 87, 126
Rhine, 18, 19, 22, 30, 31, 34, 89, 122, 123
ruling, 41, 60

saving power, 66, 86, 126
science/natural science, 2, 4, 8, 9, 11, 17, 25, 30, 33, 37, 39, 40, 55–58, 61, 68, 69, 73, 75, 76, 80, 88, 89, 95, 101, 108, 110, 111, 117–19, 122
Science Wars, 39, 40
Scientific Revolution, 117–19

self-reflection, 60
semblance (*Schein*), 99
sky, 70, 87, 89, 121, 124
slumber, 7, 70, 118
social, 44, 47, 54, 57, 58, 69, 91, 113
sociology, 2, 3, 30, 51, 55, 56, 58, 88
soul, 53, 54
space, 14–16, 24, 45, 66–68, 70–72, 74, 82–84, 95, 115, 117, 121, 122, 124
speed bump, 15, 20, 22
Spinoza, 54
standing-reserve, 17, 18, 20, 21, 34, 37, 41, 85
supreme danger, 22, 24, 88
symmetry, 4, 9, 14, 16, 19, 27, 42–44, 55, 59, 62. *See also* Asymmetry

technical mediation, 6, 13–21, 23–25, 27, 113
  combination, 14, 15
  crossing the line, 15, 16, 24. *See also* Interference
  folding, 14, 24
technique, 9, 15, 20, 89
technology, 1, 2, 4, 6–8, 12–26, 29, 30, 32–37, 41, 47, 51, 62, 65–67, 73, 76, 78, 79, 82, 83, 88, 89, 121, 123, 126, 127
temple, 36
theoretical fieldwork, 8–10, 19, 25, 29, 92, 111, 126
they, the, 47–50, 59, 60
thing, the, 7, 66, 67, 69–75, 77, 79, 82, 87, 88
thinking, 2, 4, 5, 7, 8, 18, 19, 29, 30, 38, 48, 54, 59, 66, 71, 73, 75, 77–79, 81, 82, 84, 86–88, 94, 97, 101, 109, 116, 117, 119, 120, 123, 126
Time/temporality, 3, 8, 10, 11, 14–20, 24–27, 32, 33, 35, 40, 43, 44, 46, 51, 54, 58, 60, 62–64, 67–72, 82, 85, 87, 89, 93–95, 97, 98, 101–3, 105, 106, 108, 115, 116, 119, 122–24, 126. *See also* Space

tool, 14, 19, 22, 31, 87, 112. *See also* Equipment
tradition, 57, 65, 94, 97–99, 103, 104, 112
transcendence, 56, 57, 59
transformation, 9, 24, 53, 55, 61, 115
translation, 32, 40, 53, 73, 81, 110
truth, 4, 5, 22, 33–36, 38, 39, 41, 42, 50, 79, 99, 104, 105, 108, 109
turn/turning, 6, 8, 19, 24, 50, 68, 81, 93, 96, 100, 127

uncover/uncovering/uncovered/unconcealment/unconcealing, 6, 9, 17, 44, 45, 48, 65, 84, 86, 126
understanding, 3, 4, 6, 7, 13, 16, 17, 19, 22, 24, 30, 33–37, 43, 49, 52–56, 63, 65, 66, 69, 71, 72, 75, 77, 79–81, 84–87, 93–97, 99, 100, 102, 104, 107, 110, 116–18, 122

unity, 70, 72
using, 17–19, 27, 44, 56, 57, 66, 69, 81, 107

values, 7, 21, 77, 97, 104, 105, 107, 108, 110, 111, 116

work, 3, 5–7, 9–11, 14–17, 22, 23, 25, 26, 31–42, 51, 53, 62, 72, 84, 89, 91, 95, 98, 100–102, 104, 107, 110, 112, 120–22
worker, 2, 14
world, 4, 5, 7, 15, 17–20, 22, 24–26, 29–33, 35, 36, 38, 39, 41–47, 49, 50, 52–59, 61, 67, 68, 71–75, 81, 84–88, 92, 96, 101–5, 108–11, 116, 118, 120, 122, 123, 125, 127
Wright, Frank Lloyd, 125. *See also* architecture

About the Author

Søren Riis works at the intersection of continental philosophy, philosophy of technology, and Science and Technology Studies (STS). He completed his doctoral studies in Freiburg, Germany, where the influences of Edmund Husserl and Martin Heidegger are still felt in the philosophy department. Today, he is an associate professor and director of studies at Roskilde University in Denmark. Trying to grasp some of the defining features of our time and how they are associated with technology development has been a key focus of his research for many years, and this book is the fruit of this work. Over the years, Riis has published in a number of different journals, such as *Continental Philosophy Review*, *Social Studies of Science*, *Foundations of Science*, and *Heidegger-Jahrbuch*, and coauthored the Oxford Bibliographies Online Entry on Philosophy of Technology. In his career, he has been fortunate to be a visiting scholar at Harvard University, Sciences Po in Paris, Exeter University, Stony Brook University, and the University of Twente; he also enjoyed his time as a DAAD Research Fellow at the State University of Rio de Janeiro. To encourage productive scholarly exchange and advance the kind of research he thinks is necessary, he is also a member of the board of the Danish Society of Philosophy and Danish STS.

www.ingramcontent.com/pod-product-compliance
Lightning Source LLC
LaVergne TN
LVHW051818060925
820435LV00002B/21